OXFORD PHILOSOPHICAL MONOGRAPHS

Editorial Committee

WITTGENSTEIN, FINITISM, AND THE FOUNDATIONS OF MATHEMATICS

ALSO PUBLISHED IN THE SERIES

The Justification of Science and the Rationality of Religious Belief
Michael Banner

Individualism in Social Science: Forms and Limits of a Methodology
Rajeev Bhargava

Causality, Interpretation, and the Mind
William Child

The Kantian Sublime: From Morality to Art
Paul Crowther

Kant's Theory of Imagination
Sarah Gibbons

Determinism, Blameworthiness, and Deprivation
Martha Klein

Projective Probability
James Logue

Understanding Pictures
Dominic Lopes

False Consciousness
Denise Meyerson

Truth and the End of Inquiry: A Peircean Account of Truth
C. J. Misak

The Good and the True
Michael Morris

Nietzsche and Metaphysics
Peter Poellner

The Ontology of Mind: Events, Processes, and States
Helen Steward

Things that Happen because They Should: A Teleological Approach to Action
Rowland Stout

Wittgenstein, Finitism, and the Foundations of Mathematics

MATHIEU MARION

CLARENDON PRESS · OXFORD
1998

Oxford University Press, Great Clarendon Street, Oxford OX2 6DP
Oxford New York
Athens Auckland Bangkok Bogotá Buenos Aires Calcutta
Cape Town Chennai Dar es Salaam Delhi Florence Hong Kong Istanbul
Karachi Kuala Lumpur Madrid Melbourne Mexico City Mumbai
Nairobi Paris São Paulo Singapore Taipei Tokyo Toronto Warsaw
and associated companies in
Berlin Ibadan

Oxford is a registered trade mark of Oxford University Press

Published in the United States
by Oxford University Press Inc., New York

First published 1998

British Library Cataloguing in Publication Data
Data available

Library of Congress Cataloging in Publication Data
Marion, Mathieu, 1962–
Wittgenstein, finitism, and the foundations of mathematics /
Mathieu Marion.
(Oxford philosophical monographs)
Includes bibliographical references.
1. Mathematics—Philosophy. 2. Wittgenstein, Ludwig, 1889–1951.
I. Title. II. Series.
QA8.4.M37 1998 510'.1—dc21 98–16711
ISBN 0–19–823516–X

1 3 5 7 9 10 8 6 4 2

Typeset by Hope Services (Abingdon) Ltd.
Printed in Great Britain
on acid-free paper by
Biddles Ltd.,
Guildford & King's Lynn

À Hélène, Gilbert, Nicolas, Jacques

PREFACE

During his life, Wittgenstein published only one book, the *Tractatus Logico-Philosophicus* (1921), one paper, 'Some Remarks on Logical Form' (1929), one page-long book review (1913), and one letter to the editor (1930). But he left behind an extensive *Nachlass*. From the amount of secondary literature devoted to it, it is surprising to discover that Wittgenstein wrote more on philosophy of mathematics than on any other subject. That Wittgenstein's remarks on mathematics did not receive their fair share of attention is not to be explained only by the fact that it is a comparatively unpopular topic. When Wittgenstein's *Remarks on the Foundations of Mathematics* (*RFM*) were first published in 1956, reviewers' assessments were negative. For example, the logician Georg Kreisel, a close friend who had frequent discussions with Wittgenstein in the early 1940s, ended his review of the book with these much-quoted words: 'it seems to me to be a surprisingly insignificant product of a sparkling mind' (Kreisel 1958a: 158). Since specialists were in agreement in their negative assessment, followers and commentators of Wittgenstein simply hived off issues in philosophy of mathematics from those concerning language and psychology, more or less assuming that, although Wittgenstein may have erred when tackling issues in mathematical logic and foundations of mathematics, this was of no consequence to the rest of his philosophy: they could then continue with the business of interpreting these other parts in isolation. To my mind, this is ultimately as unacceptable as it would be for someone to interpret Frege's or Russell's philosophy in ignorance of their work in mathematical logic and the foundations of mathematics.

A variety of factors explain the negative reception of Wittgenstein's remarks on the philosophy of mathematics in the 1950s. First, the book comprised remarks selected by the editors (G. E. M. Anscombe, R. Rhees, and G. H. von Wright) from various manuscripts dating from 1937 to 1944 and it is clear that editorial choices, which produced a truncated version of the original text, have hindered rather than helped us to understand Wittgenstein's thoughts. He was an undoubtedly difficult writer, but when whole paragraphs are cut out in between two published remarks without any indications to that effect, the reader may be understandably puzzled by seemingly bizarre jumps in Wittgenstein's thought. Secondly, access to Wittgenstein's writings was very limited at the time: of the posthumous writings, only the *Philosophical Investigations* (*PI*) and *The Blue and Brown Books* (*BB*) had been published. Wittgenstein's remarks, when they did not relate directly to *Philosophical Investigations*, published a few years earlier, often seemed to be coming out of the blue. One had to wait almost a decade for the first major work of the transitional period (from Wittgenstein's return to Cambridge in 1929 to the first draft of the *Blue Book* in 1933), where the later philosophy of mathematics has its source, to be published. This was *Philosophical Remarks* (*PR*) in 1964, the English translation appearing in 1975. Thirdly, Wittgenstein scholarship was simply in its infancy: it was difficult to get a good grasp of Wittgenstein's later philosophy, let alone make sense of what he had to say on mathematics. Finally, the intellectual climate of the late 1950s did not make for an easy reception of Wittgenstein's ideas: results such as Gödel's on incompleteness, Tarski's on truth, and Gödel's and Cohen's on the independence of the Axiom of Choice and the Continuum Hypothesis from Zermelo-Fraenkel's set theory had given much credence to radically non-constructive forms of Platonism in philosophy of mathematics, against which Wittgenstein's remarks seem prima facie directed. To make a long story short, if the picture of Wittgenstein presented in this book is on the whole on the right lines, the ideology of the 1950s was simply very much against Wittgenstein. For all these reasons, it is not surprising that those who read him easily misrepresented his views and rejected them flatly.

This early negative assessment of Wittgenstein's philosophy of mathematics and the concomitant idea that this part of his philosophical output is at any rate of little interest have become firmly implanted. But the situation is changing. To begin with, many more of Wittgenstein's posthumous writings are now easily available in print, and there are now shelves full of commentaries. We can safely say that our understanding of his thought has improved markedly, thanks to the work of commentators such as Gordon Baker and Peter Hacker, Anthony Kenny, Brian McGuinness, and David Pears. Moreover, the focus in mathematical logic has recently shifted considerably towards constructivism, with the growing interest in connections with theoretical computer science (especially with complexity theory): logicians are now interested not only in constructive proofs but in an even more restricted class of proofs, those that are said to be 'feasible' because they provide an algorithm which is polynomial-time computable, so that the proof can be run on a computer. In this context, issues relating to finitism and the strict finitist philosophy of mathematics, to which the name of Wittgenstein was linked by the reviewers mentioned above, have resurfaced. It is thus time slowly to piece together the various strands of Wittgenstein's philosophy of mathematics and aim for a reassessment.

One thing, however, needs to change. As pointed out, only some of Wittgenstein's later works had been published in the 1950s: the *Philosophical Investigations* in 1953, the *Remarks on the Foundations of Mathematics* in 1956, and the *Blue and Brown Books* in 1958. In postwar Oxford in particular, this editorial policy led to the development of interpretations where the later Wittgenstein, with his strategic affinities with ordinary language philosophy, was favourably contrasted with the earlier Wittgenstein, who was seen to be going in the (terribly wrong) direction of logical positivism. Thus a distorting emphasis was put on Wittgenstein II versus Wittgenstein I; and between the two great philosophies there seems to have been that grey zone called the 'transitional' period, which was considered not to contain any idea worth a closer look. Such prejudices are still very strong today, especially among those whose ideas about Wittgenstein were formed in the 1950s. True, it is better exegetically speaking to look at Wittgenstein's arguments as being directed against views held by his former self. He himself tells us in the preface to the *Philosophical Investigations* to read that work in conjunction with his *Tractatus Logico-Philosophicus* in order that we see the mistakes in the latter. But in no way does this justify ignorance of the transitional writings. To take a personal example, I began studying Wittgenstein by struggling with the *Remarks on the Foundations of Mathematics* (this time with a new enlarged edition, which still, inexplicably, does not contain the full version of the original texts) and with the topic of the day in the 1980s, the rule-following argument. I had in front of me an immense literature where almost no one cared about the development of Wittgenstein's thoughts. Hardly anyone tried to find out what it was exactly in his earlier views that he was reacting against or how his arguments took shape: it is as if Wittgenstein had a conversion. What was needed instead was a reading of his transitional writings which would provide us with an appropriate understanding of the background to his remarks on rule-following. I personally find it wrong-headed to approach Wittgenstein's arguments (as is often the case) with the idea that they are highly generic, that they are aiming at a vaguely defined *Urbild*, as if, provided that one finds the arguments convincing, a whole host of theories sharing this *Urbild* will be magically refuted or fatally undermined. It is better to try and discover the precise conception that Wittgenstein once held and that he was reacting against in the *Philosophical Investigations*. It is striking that almost one no searched for this. (Although I do not wish to present in this book a full account of Wittgenstein's remarks on rule-following, I shall say something about them in sections 6.1 and 8.1, to the effect that they have their root in the rejection of the phenomenology which is presented in Chapter 5.)

To take another example, it is even more striking that no one in the past fifty years has apparently tried seriously to understand what Wittgenstein said on mathematics in the *Tractatus Logico-Philosophicus*. A recent book by Lello Frascolla, *Wittgenstein's Philosophy of Mathematics*, contains the first adequate explanation of Wittgenstein's definition of natural numbers contained in its paragraphs 6.02–3 (Frascolla 1994: ch. 1). It is not as if one can, once again, ignore these paragraphs: not to see the importance of that topic in the overall system of the *Tractatus Logico-Philosophicus* is to be incredibly short-sighted. How can one pretend to understand Wittgenstein's later conversion if one doesn't even understand what the *Tractatus Logico-Philosophicus* is all about? Or if one does not even pay attention to the transitional writings where Wittgenstein's own arguments against his former self are formed? In this book, I have tried to weave together a few strands of Wittgenstein's philosophy of mathematics, carefully tracing the ideas back to the *Tractatus Logico-Philosophicus* and plotting their evolution through the transitional and, in some cases, later periods. This seems to me to be plain exegetical common sense.

Another cause of misunderstanding of Wittgenstein's philosophy of mathematics is an incomplete understanding of foundational debates during the first half of the century. It is not just that we are still learning about the likes of Cantor, Frege, Dedekind, Russell, and Hilbert, but also that most Wittgenstein scholars are largely ignorant of these authors, as if it is a matter of little importance to know more about the issues Wittgenstein was supposedly struggling with than an undergraduate might know. (Or it is perhaps already conceded that Wittgenstein had only a faint understanding of the issues, that he was not addressing the same ones, etc.) On the contrary, I believe that it is very important to devote large sections of the book to the explanation of issues and the various standpoints which constitute the historical context of Wittgenstein. It is only against a better-defined backdrop, as opposed to the broad sketches (not to say clichés) one usually finds, that Wittgenstein's position will be seen in sharper focus. So much of the explanatory work has to be done by carefully setting the backdrop. I shall, however, refrain as far as possible from evaluating Wittgenstein's remarks, once I have achieved some measure of understanding. I have not always succeeded in this, but I hope that my cursory evaluations, which are sometimes negative, should not be seen as the expression of a negative attitude towards Wittgenstein, but rather as signposts for further studies.

In this book I do not propose an all-encompassing, well-polished interpretation of Wittgenstein. To begin with, it is impossible to deal adequately in one book with all issues tackled by Wittgenstein in the whole of his work. Moreover, I believed, for the reasons expressed above, that a fresh start was needed, and I sought to avoid the Procrustean bed of a ready-made interpretation. I tried, rather, to follow E. M. Forster's phrase, used aptly by M. B. and J. Hintikka as an epigraph to their book *Investigating Wittgenstein* (1986): 'Only connect.' The first connection I was able to make was between Wittgenstein's remarks on quantification in 1929 and similar views of Ramsey and Weyl (see Chapter 4). The rest of my investigations were done around it; I limited myself to some central issues in the *Tractatus Logico-Philosophicus* and in writings of the transitional period. This means that I shall make only passing references to topics discussed by Wittgenstein only or mostly in his later writings. (For example, his remarks about Gödel's incompleteness theorems will not be discussed, with the exception of one footnote at the end of Chapter 1, and his remarks on the problem of consistency and the related criticisms of Hilbert's programme will only briefly be mentioned in Chapter 1. I shall, however, discuss from the *Remarks on the Foundations of Mathematics* the remarks on mathematical existence (section 6.2) and the remarks on Cantor's diagonal (7.1), and Chapter 8 will be devoted to the remarks on surveyability.)

It was neither possible nor desirable entirely to avoid discussing Wittgenstein's later remarks on mathematics. The reader whose interest lies in the later Wittgenstein should be warned, however, that I do not present a systematic, exhaustive interpretation of the later Wittgenstein. Nevertheless, I say much about it as I expand on issues dealt with in earlier remarks, especially in Chapters 6 and 8. The reader not sharing my interest in issues raised in earlier writings is thus advised to skip at least the first three chapters, which deal mainly with issues raised in the *Tractatus Logico-Philosophicus*, and perhaps come back to them at a later stage. Chapters 4 and 5 contain the core of my reading of the transitional writings. They lay out, along with the first three chapters, the groundwork for my interpretation of the later Wittgenstein, and they may profitably be read before proceeding to Chapters 6 and 8. As for Chapter 7, it may also be skipped at the outset by those whose interest is not mainly mathematical.

One difficult but fundamental topic into which I could not go in depth is that of the relationship between philosophy and mathematics. In a remark from the *Big Typescript* of 1933 which found its way in the *Philosophical Investigations*, Wittgenstein says that philosophy can only describe and not provide foundations, and that 'it leaves mathematics as it is' [*PI*, § 124]. A huge literature is based on this remark and a few others, such as those where he claims that there should be no theses in philosophy. These remarks have been interpreted with various degrees of radicalism, from simply proscribing any interpretation of Wittgenstein implying that he might have favoured a revision of current mathematics to the full oxymoron of the 'no-position position'. I have little inclination for the endless subtleties needed to make sense of the latter, but it was impossible to avoid the issue entirely, and I limited myself to some remarks on revisionism, on Wittgenstein's idea that Russell's system in *Principia Mathematica* could not be a foundational 'theory' (Chapter 1), and on his idea that there should be no 'opinion' in these matters (section 6.3). The view has been expressed that it is not possible first to explain Wittgenstein's views about logical and foundational matters and only after that to compare those views with Wittgenstein's view that there should be no theory or even theses in philosophy, because it would make his views about these matters dependent on philosophical doctrines. I believe, however, that it is possible to understand a lot about Wittgenstein's views about foundational matters without pigeonholing him; it suffices to be careful—it is as simple as that—and not to be automatically dismissive of the use of labels such as 'finitism'. (One very good example of a careful study of Wittgenstein, which influenced my reading a great deal, is Jacques Bouveresse's study (1987) of 'conventionalist' aspects of Wittgenstein's philosophy of mathematics in *La Force de la règle: Wittgenstein et l'invention de la nécessité*.) When Wittgenstein writes, for example, '[m]y theory should culminate in this that there should be no infinite proposition' (*WA* I, p. 80), I do not see how he could be said not to have held the thesis that there should be no infinite proposition, nor do I see that he did not provide any argument to that effect elsewhere in his writings. Furthermore, I fail to see why he should not be made accountable for the clear-cut implications of any thesis he held. Wittgenstein's remarks on the nature of philosophy are very programmatic, and I believe that they will be invested with meaning only after more specific remarks have been properly understood. There is definitely a danger in interpreting these remarks in isolation in such a way as to render the interpretation of the body of his text impossible. My strategy has been simply to read Wittgenstein's remarks *at face value*, to see what their implications were, and to compare the latter with various foundational stances. I think that the results will speak for themselves. To use an expression taken from Lello Frascolla, I offer them to the reader as no more than a series of 'interpretative conjectures', which I hope will be the basis for further discussion. I can only challenge critics not to repeat the same *spiel* about Wittgenstein's claim that there should be no theses in philosophy, etc., but to come up with an alternative reading of *all* the remarks quoted in this book, which will show that the many positive claims such as the one quoted above were not really meant to express theses.

> *si quid novisti rectius istis,*
> *candidus imperti; si non, his utere mecum.*

ACKNOWLEDGEMENTS

This book is a completely revised and expanded version of my D.Phil. thesis at Oxford. It has been many years in the making, and I have had the privilege over these years of meeting many first-rate minds, with whom I discussed various aspects of it. I should begin by singling out my thesis supervisor, Michael Dummett, to whom I give my warmest thanks. Although the prospect of submitting to him half-baked ideas was daunting, tutorials with him always proved to be enjoyable affairs, even when his comments were devastating. He showed much kindness, and I am deeply grateful for his help over the years. Special thanks also go to Angus Macintyre of the Mathematical Institute, who was my tutor for the academic year 1988–9, and Crispin Wright, who was also my tutor for Whitsunday term, 1988, at the University of St Andrews. As a post-doctoral fellow I had the privilege to study with him again in 1991–2, this time as a Visiting Scholar. I also studied with Jaakko Hintikka when I was a Research Fellow at the Center for the Philosophy and History of Science at Boston University in 1992–3, and with Yvon Gauthier at the Université de Montréal in 1993–4. Daniel Isaacson has been kind enough to read and comment on countless drafts of my thesis—acting as a kind of unofficial supervisor—and of this book over the years; he also deserves my warmest thanks. My intellectual debt to all these will be obvious. I must also acknowledge a particular debt over the years to four other individuals with whom I had numerous very fruitful conversations: Burton Dreben, Lello Frascolla, Juliet Floyd, and Michael Wrigley. The last two hold the promise of a further pair of excellent books on Wittgenstein's philosophy of mathematics. The following also made precious comments, for which they are warmly thanked: Alice Ambrose, George Boolos, Michael Campbell, Roy Dyckhoff, Robin Gandy, Michael Glanzberg, Hanjo Glock, Bob Hale, Peter Hacker, Edward Harcourt, Richard Heck, Georg Kreisel, Ulrich Majer, John Mayberry, Alva Noë, Alex Oliver, Charles Parsons, Michael Potter, Elizabeth Rigal, Nils-Eric Sahlin, Sanford Shieh, Robert Sinclair, Peter Sullivan, Göran Sundholm. It should go without saying that I am solely responsible for all remaining blemishes. First-rate works of scholarship have appeared in the past years; I certainly had the good fortune of studying carefully and discussing with its author Lello Frascolla's book *Wittgenstein's Philosophy of Mathematics*, already mentioned; but I deeply regret not to have been able to make use of David Stern's *Wittgenstein on Mind and Language* (1994), the content of which is closely related to that of Chapter 5. I hope to benefit from it in future work.

Without adequate financial support, it would have been impossible for me to study at Oxford and to survive during the three years between the end of my studies and my first job. I am thus grateful both to the Fonds pour la Formation de Chercheurs et l'Aide à la Recherche and to the Social Sciences and Humanities Research Council of Canada for awarding me *both* doctoral *and* post-doctoral fellowships. My research is also currently funded by a research grant from the Social Sciences and Humanities Research Council of Canada. In our increasingly philistine world, funding for governmental institutions such as these, along with funding for higher education in general, is steadily shrinking. I am perfectly aware of the privilege I had of being one of the last students in my country to make good use of the opportunity given to me, and it is with deep regret that I now see first-rate students unable to fulfil their potential because of lack of funding. While at Oxford, I also received funding from the Overseas Research Student Award Scheme and from the Cooper Fund, for which I am also grateful. I would also like to thank the Faculty of Arts and the Graduate School of the University of Ottawa for their financial support. Above all, however, it is the generous support of my family which made it possible. To them this work is dedicated, a small token of my gratitude.

CONTENTS

LIST OF ABBREVIATIONS

Wittgenstein

AWL	*Wittgenstein's Lectures, Cambridge 1932–1935*, from the notes of A. Ambrose and M. McDonald (Oxford: Blackwell, 1979).
BB	*The Blue and Brown Books* (Oxford: Blackwell 1958).
LA	*Lectures and Conversations on Aesthetics, Psychology, and Religious Belief*, from the notes of R. Rhees, Y. Smythies, and J. Taylor (Berkeley: University of California Press, 1967).
LFM	*Wittgenstein's Lectures on the Foundations of Mathematics, Cambridge 1939*, from the notes of R. Bosanquet, N. Malcolm, R. Rhees, and Y. Smythie, (Ithaca, NY: Cornell University Press, 1976).
LO	*Letters to C. K. Ogden with Comments on the English translation of the 'Tractatus-Logico-Philosophicus'* (Oxford: Blackwell, 1973).
LWL	*Wittgenstein's Lectures, Cambridge 1930-1932*, from the notes of J. King and D. Lee (Oxford: Blackwell, 1980).
M	'Wittgenstein's Lectures in 1930–3', in G. E. Moore, *Philosophical Papers* (London: Allen & Unwin, 1959), pp. 252–324.
NB	*Notebooks 1914–1916*, 2nd edn. (Oxford: Blackwell, 1979).
OC	*On Certainty* (Oxford: Blackwell, 1969).
PG	*Philosophical Grammar* (Oxford: Blackwell, 1974).
PI	*Philosophical Investigations* (Oxford: Blackwell, 1953).
PR	*Philosophical Remarks* (Oxford: Blackwell, 1975).
RC	*Remarks on Colour* (Oxford: Blackwell, 1977).
RFM	*Remarks on the Foundations of Mathematics*, rev. edn. (Oxford: Blackwell, 1978).
SRLF	'Some Remarks on Logical Form', *Proceedings of the Aristotelian Society, Supplementary Volume 9* (1929), 162–71.

TLP — *Tractatus Logico-Philosophicus*, introduction by Bertrand Russell (London: Routledge & Kegan Paul, 1922).

WA i — *Wiener Ausgabe*, i: *Philosophische Bemerkungen* (Vienna: Springer, 1994).

WA ii — *Wiener Ausgabe*, ii: *Philosophische Betrachtungen*, *Philosophische Bemerkungen* (Vienna: Springer, 1994).

WVC — *Ludwig Wittgenstein and the Vienna Circle*, from the notes of F. Waismann (Oxford: Blackwell, 1979).

Z — *Zettel* (Oxford: Basil Blackwell, 1967).

References to Wittgenstein's unpublished manuscripts and typescripts follow G. H. von Wright's catalogue in 'The Wittgenstein Papers', in L Wittgenstein, *Philosophical Occasions: 1912–1951*, ed. J. Klagge and A. Nordmann (Indianapolis: Hackett, 1993), pp. 480–510.

Whitehead and Russell

References to A. N. Whitehead and B. Russell, *Principia Mathematica* are to the paragraph (*) number or to the volume and page number of the second edition:

PM — *Principia Mathematica*, 2nd edn. (3 vols., Cambridge: Cambridge University Press, 1925–7).

1

Introduction: Wittgenstein's Anti-Platonism

> Faire de l'algèbre, c'est *essentiellement* calculer.
>
> Bourbaki

Constructivism in mathematics is usually said to have its origin in the reaction of Leopold Kronecker, the great nineteenth-century German mathematician, to the rise of Cantor's set theory. This is in a sense misleading: Kronecker was not so much the first constructivist as the greatest representative at the time of a tradition of algebraists which includes Newton, Tchirnhausen, Fermat, Euler, Jacobi, Sturm, Sylvester, Kummer—all responsible, as Kronecker was, for the discovery of powerful algorithms—and which has continued since with names such as Hurwitz, König, and Macaulay. One is reminded here of what Shreeram Abhyankar has dubbed 'high-school algebra': algebra as it was before the advent of the set-theoretical 'modern algebra'. It dealt with systems of polynomial equations and power series, a principal aspect of which being elimination theory and its algorithmic procedures.[1] Kronecker obtained deep results in arithmetic of modular functions and the theory of elliptic functions with complex multiplication which turned out to be of fundamental importance for twentieth-century mathematics.[2] This fact is known to mathematicians, but in

[1] For more details, see Abhyankar's 'Historical Ramblings in Algebraic Geometry and related Algebra' (1976). Elimination theory is a subject to which Kronecker made substantial contributions. It is fitting to note that this theory is at the basis of the model-theoretic theory of quantifier elimination. On the latter, see Dickman (1992), Macintyre (1986), Sinaceur (1991), and especially Lou van den Dries' 'Alfred Tarski's Elimination Theory for Real Closed Fields', where its Kroneckerian nature is explicitly recognized (van den Dries 1988: 11). On another note, the algebraic viewpoint alluded to here differs from and is much more fruitful than a more traditional, constructivist foundational standpoint such as intuitionism, as can be seen from Kreisel and Macintyre (1982).

[2] For the relevance of Kronecker's seminal results on modular and elliptic functions for contemporary mathematics, see André Weil's *Elliptic Functions according to Eisenstein and Kronecker* (1976) and, especially, the third part of Serge Vladut's

the tradition of logical foundations[3] Kronecker has been known and universally condemned as the villain who committed the unpardonable sin of opposing Georg Cantor's pioneering work in set theory. For example, David Hilbert, the doyen of German mathematics at the turn of the century, caricatured him as a *Verbotsdiktator* who opposed set theory on purely dogmatic philosophical grounds (Hilbert 1935: 159, 161). But the surprisingly insignificant number of remarks of a purely philosophical nature in the whole of Kronecker's writings—a few paragraphs and footnotes—simply points to a lack of interest in philosophical or foundational questions. Indeed, Kronecker admitted in a letter to Cantor that he assigned to these questions only a 'secondary value' (Meschowski 1967: 238). In the same letter he compared himself to his teacher and close friend, Ernst Kummer: 'I recognized, as he did, the unreliability of all speculations and I took refuge in the safe haven of real mathematics', adding a little bit later:

> In the field of mathematics, I find a real scientific value only in concrete mathematical truths or, to put it more pointedly, 'only in mathematical formulas'. History of mathematics has shown that only these are everlasting. The various theories about the foundations of mathematics (such as that of Lagrange) were put aside over time, but Lagrange's resolvent is here to stay! (pp. 238–9)

What Kronecker called 'mathematical formulas' in his letter we would today call 'algorithms'. His words should thus come as no surprise, considering the importance of algorithms in Kronecker's mathematical practice. It could be argued that Cantorian set theory cannot be dissociated from a Platonist philosophy, that is that it is essential to Cantor that, in parallel to our intensional understanding of a concept such as that of natural numbers, its extension is already to exist in its totality. While Cantor has devoted much space to this philosophical question, trying to overturn the then prevalent Aristotelian view of infinity according to which there are no completed infinities,

Kronecker's Jugendtraum and Modular functions (1991). Harold Edwards has also developed a theory of divisors on the basis of Kronecker's own theory (which is found in *Grundzüge einer arithmetischen Theorie der algebraischen Grössen* (Kronecker 1899: ii. 287–96)) in *Divisor Theory* (Edwards 1990), which shows its relevance for modern linear algebra. An early introduction to Kronecker's algebraic number theory is Hermann Weyl's *Algebraic Theory of Numbers* (1940).

[3] I am using the expression 'logical foundations' in a very broad sense, derived from Kreisel, for whom the rival programmes associated with Zermelo (set theory), Hilbert (formalism), and Brouwer (intuitionism) are in fact just variants aiming at correcting, simplifying, refining, and extending Frege's scheme. See Kreisel (1984).

Kronecker, who did not see any need in his work for Cantor's new concepts, had simply no interest in these questions.

Reading Kronecker's letter to Cantor, one cannot be but reminded of Wittgenstein:

> It is a strange mistake of some mathematicians to believe that something inside mathematics might be dropped because of a critique of the foundations. Some mathematicians have the right instinct: once we have calculated something it cannot drop out and disappear! And in fact, what is caused to disappear by such a critique are names and allusions that occur in the calculus. (*WVC*, p. 149)

Wittgenstein set forth in his writings of the transitional period a distinction between 'calculus' (*Kalkül*) and 'prose' (*Prosa*), which has received some attention in the secondary literature.[4] But commentators have usually failed to notice that by 'calculus' Wittgenstein also meant 'algorithms'. There is no doubt that he conceived of mathematics as being essentially algorithmic, that is as some sort of high-level abacus activity. This fundamental feature of Wittgenstein's philosophy of mathematics is in line with Kronecker's 'philosophy' (although Wittgenstein practically never referred to Kronecker and probably never read his work).[5] We shall see in the next chapter that Wittgenstein had already espoused this 'algorithmic' viewpoint in his first major work, the *Tractatus Logico-Philosophicus*; his attempt at a form of reduction of arithmetic to a theory of 'operations' (as opposed to a theory of 'classes' such as that of Russell's *Principia Mathematica*) making very little sense otherwise. It is also worth reflecting here on the nature of Wittgenstein's major contribution to logic in that very book: the development of truth-tables as a *decision procedure*.

This 'algorithmic' viewpoint was also at the forefront during the transitional period. Waismann reported Wittgenstein as saying:

[4] Stuart Shanker, for example, makes good use of this distinction in his *Wittgenstein and the Turning-Point in the Philosophy of Mathematics* (1987).

[5] That Wittgenstein probably never read Kronecker is not difficult to establish, since very few mathematicians (and probably no philosophers) read his works in this century. The only reference to Kronecker in Wittgenstein's writings that I know of is found in the *Big Typescript* (TS 213, pp. 497–8), where he points out that Kronecker should not have said 'Die ganzen Zahlen hat der liebe Gott gemacht, alles andere ist Menschenwerk' but rather 'nur die Kardinalzahlen sind wirkliche Zahlen'. Kronecker's remark is not to be found in any of his writings; it is cited by Heinrich Weber in his obituary to Kronecker, where he claimed that the latter uttered it in a lecture given in Berlin in 1886 (Weber 1893: 19). These words have since become *the* statement of Kronecker's philosophy for those who have never read him.

> I believe that mathematics, once the conflict about its foundations has come to an end, will look just as it does in elementary school where the abacus is used. The way of doing mathematics in elementary school is absolutely strict and exact. It need not be improved upon in any way. Mathematics is always a machine, a calculus. The calculus does not describe anything. (*WVC*, p. 106)

A little bit further on in the conversation, Wittgenstein added: 'A calculus is an abacus, a calculator, a calculating machine; it works by means of strokes, numerals, etc.' (p. 106). (One is reminded here of the literal meaning of the older German expression for algebra, *Buchstabenrechnung*.) He also wrote in the *Philosophical Grammar*:

> Mathematics consists entirely of calculations.
>
> In mathematics *everything* is algorithm and *nothing* is meaning; even when it doesn't look like that because we seem to be using *words* to talk *about* mathematical things. Even these words are used to construct an algorithm. (*PG*, p. 468)

In order to appreciate Wittgenstein's analogy between mathematics and the abacus, we need to reflect a moment on his use of analogies. It is well known that Wittgenstein often compared mathematical propositions to rules of grammar; he claimed, for example, that arithmetic *is* the grammar of numbers. It is obvious, however, that the claim was not that arithmetical propositions were rules of grammar *stricto sensu*, since this would involve a controversial extension of the concept of a 'rule of grammar'.[6] The claim was made in order to underline a fundamental dimension of the use of arithmetical propositions. The image of the Russian abacus is used in a similar fashion, and when Wittgenstein claimed that the signs in mathematics '*are* like the beads of an abacus' (*PR* § 157), he wanted to direct our attention towards an aspect of mathematics neglected within the tradition of logical foundations, which could be framed very crudely as the claim that the core of mathematics is algorithmic.[7] Many crucial pieces of

[6] G. E. Moore famously objected to Wittgenstein, in his 'Wittgenstein's Lectures in 1930–33', that he was not using the expression 'rules of grammar' in any ordinary sense (*M*, pp. 276 f.).

[7] There is an extensive literature on Wittgenstein's claim that there should be no substantial 'theses' in philosophy (see e.g. *PI*, § 128). To see Wittgenstein as defending the *thesis* that mathematics consists of algorithms seems prima facie to violate his own requirement. To settle this question would lead us too far afield, but it should be pointed out, in defence, that the thesis that mathematics is an activity consisting in providing algorithms is in many ways *trivial*—at least one could claim that Wittgenstein thought that it was so. Thus one would respect the spirit of remarks such

the puzzle fall into place if we adopt this interpretative thesis: we can begin to understand, for example, Wittgenstein's preference for quantifier-free systems of arithmetic (see Chapter 4 below), his remarks on the Law of Excluded Middle (section 6.2), and his remarks on the continuum (Chapter 7).

To say that in mathematics '*everything* is calculus and *nothing* is meaning' may seem too strong a claim: we do use words from our ordinary language in the process of doing mathematics, even while proving mathematical propositions. In fact, it is essential for proofs that they incorporate words of our ordinary, everyday language. These words are what Wittgenstein called the 'everyday prose that accompanies the calculus' (*WVC*, p. 129). The distinction between 'prose' and 'calculus' was seen by Wittgenstein as a key distinction in philosophy of mathematics:

> . . . what is caused to disappear by (a critique of foundations) are names and allusions that occur in the calculus, hence what I wish to call *prose*. It is very important to distinguish as strictly as possible between the calculus and this kind of prose. Once people have become clear about this distinction, all these questions, such as those about consistency, independence, etc., will be removed. (*WVC*, p. 149)

The appearance of prose is necessitated, according to Wittgenstein, by the fact that a mathematical proof shows us something that it cannot say by itself. Hence the need, so to speak, to express the inexpressible—this being a prime example of what Wittgenstein would call 'running against the limits of language'—and introduce everyday prose in mathematics:

> The proof lets us see something. What it shows, however, cannot be expressed by means of a proposition. Thus it is also impossible to say, 'The axioms are consistent.' (Any more more than you can say, 'There are infinitely many numbers.' That is everyday prose.) (*WVC*, p. 137)

as *PI*, § 128. Indeed, it is undeniably true that mathematical practice requires more than just existential proofs. and ultimately an algorithm for the computation of solutions. That this trivial fact is usually forgotten and seldom stressed by philosophers of mathematics is explainable by the fact that when they reflect on mathematical activity, they usually have in front of them the achievements of Cantor, Dedekind, Frege, and Hilbert. These authors have at least this in common that they fostered a style of mathematics, the 'modern axiomatic method', for which algorithmic results are too burdensome, not abstract enough, somewhat inessential. It should not be forgotten that the algorithmic approach of Kronecker became almost extinct during Wittgenstein's lifetime, as a result of the immense initial success of Cantor's set theory and Dedekind's algebraic number theory—both popularized by Hilbert.

(This extension of the showing/saying distinction of the *Tractatus Logico-Philosophicus* will be discussed in section 4.2.)

To think that mathematics consists of something more than calculations with signs means that the signs can be thought of as standing proxy for or 'describing' something. Against the tendency to think in those terms, Wittgenstein insisted repeatedly there is no such mathematical or logical activity as 'describing objects'. Rather, the signs 'do' mathematics:

> Let's remember that in mathematics, the signs themselves *do* mathematics, they don't describe it. The mathematical signs *are* like the beads of an abacus. (*PR* § 157)

> . . . we can't describe mathematics, we can only do it. (And that of itself abolishes every 'set theory'.) (*PR* § 159)

Therefore, as a corollary of his fundamental claim about the algorithmic nature of mathematics, Wittgenstein had to oppose the claim that mathematics consists in the description of pre-existing (abstract) structures. This thesis is not just a superfluous bit of philosophy; it is an integral part of some of the concepts underlying the axiomatic approach of Cantor, Hilbert, Frege, and Russell. It is the 'extensionalism' which we find in the writings of Wittgenstein's contemporaries at Cambridge in the late 1920s such as Frank Ramsey, who thought, at least at the time he wrote his paper 'The Foundations of Mathematics', that 'extensionality' is a 'fundamental characteristic of modern analysis' which was, according to him, neglected in *Principia Mathematica* (Ramsey 1978: 165). Ramsey then set out to modify the theory of types in order for it to reflect the 'extensional attitude of modern mathematics' (Ramsey 1978: 174).[8] (The clashes between Ramsey the 'extensionalist' and Wittgenstein will be detailed in sections 2.3, 3.2, and 6.3.)

This 'extensionalism' finds its source in the mutation of the notion of function during the nineteenth century. Differentiability and definability restrictions were gradually lifted, and the now common notion of an 'arbitrary function' was introduced.[9] In particular, Peter Gustav Lejeune-Dirichlet was led early in the century by a result of his on

[8] This criticism of *Principia Mathematica* was already present in Johnson's *Logic*, where the need for 'classes taken in extension' was also emphasized (Johnson 1922: 166, 174). Although Russell would ultimately agree with Ramsey and Johnson on this point, Wittgenstein took the opposite stance, rejecting the notion of infinite classes taken in extension.

[9] For some elements of this history, see Burgess (1995: 90–7).

the convergence of Fourier series[10] to introduce a general concept of a function of a real variable according to which *y* is called a function of *x* if there is, within a definite interval, a definite value of *y* for every value of the variable *x*, while it does not matter whether *y* is dependent on *x* according to the same rule within the whole interval or not, and whether the dependence can be expressed by means of mathematical operations or not (Dirichlet 1969: 135). This was the first statement of the general notion of an 'arbitrary function' which became associated with Dirichlet's name. The subsequent widespread adoption of this notion meant the abandonment of the earlier notion of function, according to which a function is 'defined by a formula'. In very crude terms, what Dirichlet observed is that existing results about functions would remain true if one replaced the traditional notion of a function 'defined by a formula' by that of its graph (i.e. the result of applying the rule stated by the 'formula').

Adopting Dirichlet's notion involved no loss in existing mathematics but gains in constructions. For example, the new approach ultimately led to an easy generalization of analysis to function spaces. Thus, with Dirichlet mathematicians moved from an intensional notion of function-as-a-rule to a purely extensional conception. Prominent in the drive to discard the older notion of function was Bernhard Riemann, whose work had, alongside Dirichlet's, a considerable influence on mathematicians of the time such as Dedekind and Weber,[11] and Cantor, who gave the first formulation of what were to become the transfinite numbers precisely while working on a problem framed by Dirichlet, that of the uniqueness of the representation by trigonometric series of arbitrary functions. This is in fact the origin of the calculus of set theory,[12] within which the notion of function as a set of ordered pairs was developed over the years. A current definition of the notion of function would now read thus: a function from, say, A to B is a subset $F \subseteq A \times B$ with the property that

[10] In 1829, Dirichlet proved that the Fourier series of a (piecewise monotone) function converges to the function. In modern terms, Dirichlet proved that the Fourier series converges pointwise to the function for any periodic function with period 2π that is piecewise monotone and has at most a finite number of discontinuities, at each of which the function has a left-hand limit and a right-hand limit and a value that is the average of these two limits.

[11] Dedekind and Weber explicitly stated in the introduction to their masterpiece, 'Theorie der algebraischen Funktionen einer Verändlichen' (1882: 238–41), where they presented the first unified treatment of algebraic number theory and algebraic curve theory, that it was Riemann's work on Abelian functions which motivated their work.

[12] See Dauben (1979: 30–46) or Hallett (1984: 3–6).

$$\forall x \in A\ \exists! y \in B < x, y > \in F$$

With this definition, functions are given as a whole in extension and the door is open to arbitrary subsets or arbitrary functions for which no computation rule (i.e. no definition by a formula) can be given.

On the opposite side stood Kronecker. One can extract from his mathematical practice three tenets which can be stated simply: first, everything must be constructed from the natural numbers. Secondly, no completed infinities are allowed. Thirdly, proofs of existence are to provide a method or algorithm to find *in a finite number of steps* an arbitrarily good approximation to the number whose existence is proven and a definition, in number theory or algebra, is acceptable only if it can be checked (again by an algorithm) in a finite number of steps whether any given number falls under it or not.[13] Kronecker is also said to have been of the opinion that an infinite series is admissible only if a rule for computing its terms is given, although there seems to be no clear statement supporting this view in his writings.[14] One finds, however, a good statement of the Kroneckerian stance on

[13] The usual basis for such a characterization is an account by Kronecker's student Kurt Hensel: 'I must also point out a requirement which Kronecker consciously imposed on the definitions and proofs of general arithmetic [*allgemeinen Arithmetik*], the strict observance of which distinguishes his treatment of number theory and algebra from almost all the others. He believed that in these domains one could and must formulate each definition in such a way that one could verify in a finite number of steps whether it applies to a given magnitude or not. Similarly, a proof of the existence of a magnitude could only be seen as completely rigorous if it contained a method by which the magnitude whose existence was being claimed could really be found. Kronecker was far from wanting to reject entirely a definition or proof which did not satisfy the highest demands; but he believed that in this case something was lacking and he held that any improvement in this direction was an important problem, through which our knowledge of an essential point could be extended. Besides, he believed that a formulation which was rigorous in that respect would take in general a simpler form than another which did not fulfil these requirements' (Kronecker 1901: vi). For a fuller discussion of Kronecker's standpoint, see Edwards (1980; 1987; 1989; 1992a; 1992b), and Marion (1995a).

[14] This much can be inferred (among other things) from a footnote to Kronecker's paper 'Über einige Anwendungen der Modulsysteme auf elementare algebraische Fragen': 'It seems to me that these considerations are in opposition to the introduction of Dedekind's concepts of "module", "ideal", etc., as well as to the introduction of various new concepts, which have been used in many recent attempts (first of all by Heine) to grasp and to give foundations to the concept of 'irrational' in all its generality. The *general* concept of an infinite series, e.g. one which increases according to definite powers of variables, is in my opinion permissible only with the reservation that, in every special case, certain assumptions must be shown to hold, on the basis of the arithmetical laws of construction of the terms (or coefficients) . . . which allow the use of the series as finite expressions and thus make it really unnecessary to go beyond the concept of a *finite* series' (Kronecker 1899: iii. 156).

arbitrary functions in Julius Molk's revised French translation (Molk 1909) of Alfred Pringsheim's contribution to the *Encyklopädie der mathematischen Wissenschaften* (Pringsheim 1899). According to Molk, who was a student of Kronecker, Dirichlet's notion would make sense only if one was able to write down an 'ideal table':

> In order to bring to the fore the arithmetical dependence of a (real) variable x and a *function* y of this variable x in a domain (x), in the general sense of the word given by G. Lejeune-Dirichlet, one would need to draw a kind of 'ideal table' in which each value of y is facing the corresponding value of x. (Molk 1909: 20)

But, apart from the case where the domain is finite, the 'ideal table' must contain an infinity of elements, and

> one cannot see how it could be effectively realized. There is no need for this since, in order to study in a precise manner the arithmetical dependence of y and of x, it is not necessary to encompass it all in one look; it suffices to obtain at any moment, in a rigorous fashion, those of the elements which are needed.
>
> This is the case when the ideal table is, so to speak, condensed in a *computational procedure* from which one obtains effectively the value of y corresponding to each value of x in the domain (x). (p. 20)

So Molk concluded his discussion by requesting that functions be given by a rule:

> when we say with G. Lejeune-Dirichlet, that a (real) variable y is a (real) *one-to-one function* of a (real) variable x, in a domain (x), when to each value of x in the domain (x) corresponds a definite value of y, we cannot dispense with the supposition that this definite value of y is, either directly or indirectly, *defined* with the help of the corresponding value of x by some *procedure of computation* . . . (p. 22)[15]

We can see here that the rejection of the notion of an arbitrary function is linked with a typically Kroneckerian emphasis on algorithms.[16] Through the influence of Alsatian students such as Molk

[15] For Pringsheim's original statement, see Pringsheim (1899: 9 f.).

[16] As an example of the opposite standpoint, I should quote the objections (to Pringsheim's original article) raised by a Cantorian from Cambridge who was a friend of Russell, Philip Jourdain: 'This implies that the function must be defined by at most an *enumerable* aggregate of specifications. However, such a restriction, which would reduce the cardinal number of all functions to be considered from $(2^{\aleph_0})$ to $(\aleph_0)$, and would in general, exclude integrable functions, is certainly not necessary for the theorems on the upper and lower limits of a function, and, in any case, is a practical necessity irrelevant to our contemplation of functions *sub specie aeternitatis*' (Jourdain 1905: 185–6 n.).

and, especially, Jules Drach, Kronecker left a profound imprint on a generation of brilliant French mathematicians at the turn of the century.[17] Émile Borel, René Baire, and Henri Lebesgue all believed that admissible objects must be defined by an expression which is a finite sequence of symbols taken from a finite alphabet—in their own words, that objects must be 'effectively defined' or defined 'in a finite number of words'.[18] This position, which we could label 'definabilism', also motivated further work of a set-theoretical nature by Russians such as Nicolas Lusin and Poles in Warsaw such as Alfred Tarski.[19] In a famous exchange of letters (to be discussed further in section 3.3) on the occasion of Zermelo's proof of the Well-Ordering Theorem, Jacques Hadamard argued against his colleagues that their definabilist viewpoint was incompatible with Cantor's results, since the number of mathematical objects definable in a finite number of words can only be countable, while by virtue of Cantor's result about the cardinality of the set of real numbers there are uncountably many objects. Thus some presumably admissible mathematical objects cannot be 'defined in a finite number of words'. Cantor's diagonal argument, which is at the heart of his results about real numbers, also played an important role in a paradox devised by another French mathematician, Jules Richard, where an object is defined which is not definable by any of a previously defined enumeration of all possible definitions in a finite number of words (Richard 1967). This paradox led Poincaré to reject as vicious circular definitions (such as the definition in terms of definability involved in Richard's paradox). As a result of the polemic with Russell and Couturat, Poincaré changed his mind—he had previously believed that existence meant simply 'freedom from contradiction', a looser criterion—and became an ardent defender of definabilism, his predicativism being in fact a radical form of the latter (see section 2.3).[20]

[17] To measure the importance of the role played by Jules Drach, see Borel (1972: 1439–40).

[18] See e.g. Borel (1914: 152, 157; 1972: 1440, 1079).

[19] Constructively motivated work on the definability theory of the continuum led, however, to the emergence of descriptive set theory. See Kanamori (1995). It is fitting to point out here that it is within the context of his investigations of definability that Tarski obtained (*c*.1927–9) what is perhaps the most significant result, mathematically speaking, of his career, that of elimination of quantifiers for real closed fields, a final proof of which he published in *A Decision Method for Elementary Algebra and Geometry* (Tarski 1951). For details, see van den Dries (1988).

[20] As we shall see in section 2.3, Hermann Weyl adopted Poincaré's predicativism in his first major work on foundations of mathematics, *The Continuum* (Weyl

At the time of Dirichlet, there seemed to be little or no reward in using more information on rules; but the development of constructive mathematics and work on computability, from Turing onwards, changed this situation. Broadly speaking, the constructivist reaction to Dirichlet's arbitrary functions led to the development within logical foundations, especially in the 1930s, of precise theories about computable functions, which reduced the vagueness of the concept of function 'defined by a formula'. (This is a major success for logical foundations.) While the theory of recursive functions immediately comes to mind, it is worth emphasizing here the invention, again in the 1930s, of a new notation for functions, and the subsequent development of various systems of lambda-calculus (Church 1932).[21] For a function 'defined by a formula', say, the function f from **R** to **R** defined by:

$$f(x) = 3x^2 + 1$$

and x a bound variable, the lambda-calculus notation will be:

$$\lambda x.\ 3x^2 + 1$$

In lambda-calculus, a term $\lambda x.$ M represents the 'operator' whose value at an argument N is calculated by substituting N for x in M. So, in the above example, for N = 2, we have:

$$(\lambda x.\ 3x^2 + 1)2 = 13$$

A typical operator is that of identity or I, which is defined thus:[22]

$$\mathrm{I} \equiv \lambda x.\ x$$

It is worth noticing that there is no equivalent for this operator in systems of set theory such as Zermelo-Fraenkel's, where each set has an identity function but there is no universal identity function that can be applied to everything. Differences of this sort are due to the

1987). One finds a direct echo of Molk in Weyl's book when he speaks of the '*completely vague* concept of function which has become canonical in analysis since Dirichlet' (Weyl 1987: 23). We shall also see in section 4.1 that Weyl 'converted' to intuitionism in 1921.

[21] The lambda notation is obviously not a trivial new one. It led to a fruitful alternative definition of the concept of a computable function, that of a λ-definable function, as is seen from the fact that the first undecidability result was obtained by Church in the lambda-calculus, and not in the theory of recursive functions.

[22] For information about lambda-calculi, the *locus classicus* is Barendregt (1984). See also Hindley and Seldin (1986).

non-set-theoretic origin of the lambda-calculus, of which we have lost sight (perhaps as a result of Dana Scott's influential analysis of computability, in the 1970s, which led to the development of a semantic model of lambda-calculus and to the theory of denotational semantics for programming languages).[23] The notion that lambda-calculus was originally devised to capture is, however, precisely not that of a function as a set of ordered pairs but, rather, that of a function as an 'operation' which may be applied to some 'object' to produce another 'object'. The latter notion was given the name of 'operator' in order to distinguish it from the set-theoretical notion of function. An essential difference between these two notions is that an operator is defined by describing how it transforms an input (i.e. by a 'formula'), without defining the set of inputs, i.e. without defining its domain. Moreover, there is no restriction on the domain of some operators. (Some operators are self-applicable, but that is not the case with functions.) To use again the terminology of intension and extension, we can say that the notion of operator is intensional, while the notion of function as a set of ordered pairs is purely extensional.[24]

That it is not entirely inappropriate to use the historical developments just described as a backdrop to our study of Wittgenstein's philosophy of mathematics will easily be seen from the next section, where it will be shown that Wittgenstein's notion of 'operation' and his definition of integers in proposition 6.02 of the *Tractatus Logico-Philosophicus* are a prefiguration of the lambda-calculus. It is also fitting to notice also that Wittgenstein was perfectly aware of the debate surrounding Dirichlet's notion of an arbitrary function. For example, he is on record as pointing out that 'set theory starts from Dirichlet's concept of a function' (*WVC*, p. 102). He also thought that Dirichlet's concept of a function of a real variable made sense only if one could produce an 'ideal table', which he here calls a 'list':

> A law is not another method of giving what a list gives. The list *cannot* give what the law gives. No list is imaginable any more. We are actually dealing with two absolutely different things. People always pretend that the one is an indirect method of doing the other. I could supply a list; but as that is too complicated or beyond my powers, I will supply a law. This sounds like saying, Up to now I have been talking to you; when I am in England I shall have to write to you. (*WVC*, p. 103)

[23] For Dana Scott's original presentation, see Scott (1970).

[24] The philosophical remarks in this passage are taken from Hindley and Seldin (1986: chs. 1 and 3).

Wittgenstein's objections both to Russell's definition of identity in *Principia Mathematica* and to Ramsey's 'functions in extension', and his remarks on the Axiom of Choice and on the notion of numerical equivalence (these topics will all be discussed in Chapter 3), make very little if any sense if one does not assume that he rejected the notion of an arbitrary function.

Use of the intension/extension terminology seems also warranted, although it leads to crude characterizations. Wittgenstein himself used sometimes the expression 'extensionalism' in order to describe the position that he was criticizing.[25] The distinction drawn by him between finite sequences and infinite series (to be discussed in section 6.3) is readily understandable in those terms: for Wittgenstein there is simply no such thing as a non-rule-governed extensional specification of an infinite set. The claim that infinite sets should only be specified intensionally helps in turn to explain his remarks on the continuum (see Chapter 7). On a more general level, the intension/extension terminology greatly helps to clarify the differences between Wittgenstein's philosophy of mathematics and Platonism, intuitionism, and strict finitism. The example of the Fibonacci sequence at the end of section 6.3 will show that Wittgenstein thought that the underlying assumption of the existence of an extension was an element of Platonism which had survived in some constructivist positions such as intuitionism and strict finitism (which, confusingly, he sometimes called 'finitism'). This helps in turn to explain Wittgenstein's criticisms of intuitionists on the mathematical existence and the Law of Excluded Middle (section 6.2) and his rejection of the argument of epistemic limitations (see section 6.3), the consequence of which is the repudiation of the cornerstone of strict finitism (section 8.1).

Instead of talking of 'anti-extensionalism', I would prefer to call Wittgenstein, in a more neutral way, an 'anti-Platonist'. There seem to be, roughly speaking, two broad types of anti-Platonism. For the first, 'moderate' form of anti-Platonism, mathematical structures may have been originally set up by us, but nevertheless acquire some 'autonomy': once we create the natural numbers series, we *discover* facts about it. Such a form of anti-Platonism was propounded by Friedrich Waismann in his lectures at Oxford, in the 1950s:

[25] e.g. in *PR* § 130, *PG*, p. 457.

We *generate* the numbers, yet we have no choice to proceed otherwise. There is already something there that *guides* us. So we make, and do not make mathematics.

. . . We cannot *control* mathematics. The creation is stronger than the creator. (Waismann 1982: 33)

Towards the end of his review of Wittgenstein's *Remarks on the Foundations of Mathematics*, Michael Dummett had also proposed, independently, against the foundational metaphors of the mathematician as a 'discoverer' or as an 'inventor', the 'intermediate picture' of 'objects springing into being in response to our probing' (Dummett 1978a: 185).

Wittgenstein should certainly be seen, as foundational metaphors go, as an anti-Platonist, since he insisted frequently on the fact that the mathematician 'is an inventor, not a discoverer' (*RFM* i, § 168).[26] As early as December 1929, Wittgenstein mentioned to Schlick and Waismann: 'What we find in books on mathematics is not a *description of something*, but the thing itself. We *make* mathematics' (*WVC*, p. 34). From this quotation and related ones, one can see how Wittgenstein's anti-Platonism is linked with his 'algorithmic' viewpoint. His brand of anti-Platonism is more extreme than that of Waismann or Dummett, since he insists that we *never* discover facts about structures that we have already set up: any new theorem is in fact a new extension of mathematics. In a later meeting, he would add that '[i]n mathematics it is just as impossible to discover anything as it is in grammar' (*WVC*, p. 63). This second, more extreme form of anti-Platonism pervades the whole of Wittgenstein's philosophy of mathematics. In particular, it underlies his criticisms of the notion of numerical equivalence, which plays an important role in the logicist definition of the natural numbers (see section 3.3). It is also related to his famous rule-following argument, which is partly directed against the Platonist imagery of 'rules-as-rails' (*PI* §§ 218–19).[27] The later is in fact acceptable to moderate forms of anti-Platonism.

[26] See also *LFM*, p. 22, and *RFM* i, § 32; app. ii, § 2 and vii, § 5. Against Russell's careless comparison of logic and mathematics with zoology (1919: 169), Wittgenstein said of mathematics that it is neither akin to zoology (*AWL*, p. 225; *RFM* iv, § 13), to mineralogy (*RFM* iv, § 11), to 'pomology' (the study of apples) (*AWL*, pp. 101–2) or to physics (*LFM*, p. 240). These are but some of the many manifestations of Wittgenstein's annoyance with the idea of mathematics as a kind of 'natural history' (*PG*, p. 369; *RFM* iv, §§ 11–13 and vi, § 49) and with the idea of logic as a kind of 'ultra-physics' (*RFM* i, § 8), this last expression originating in Hardy (1929: 18).

[27] Criticisms of Platonism implicit in Wittgenstein's rule-following argument will be discussed in section 8.1; but see also Floyd (1991) for a different approach to the subject.

This radical opposition to Platonism is absolutely fundamental; much of Wittgenstein's philosophy makes little sense without it. To give another example, we can also begin to make sense of Wittgenstein's critical remarks about Hilbert's metamathematics. Indeed, Wittgenstein always insisted that '*calculation with letters is not a theory*' and, consequently, argued over and over that 'Hilbert's "metamathematics" must turn out to be mathematics in disguise' (*WVC*, p. 136). I shall not discuss this topic further. It should be noticed, however, that precisely this point is behind Wittgenstein's annoyance with the supposed need for a consistency proof of the kind Hilbert was looking for.[28] Wittgenstein made similar remarks in his conversations with Schlick and Waismann, and later in *Philosophical Grammar* about Whitehead and Russell's system in *Principia Mathematica*. He claimed indeed that, as in the case of Hilbert's metamathematics, the prose in *Principia Mathematica* could only be a confused attempt at providing a *theory* (this being precisely the error committed by the Platonist): 'what is calculus must be separated off from what attempts to be (and of course cannot be) *theory*. The rules of the game have to be separated off from inessential statements about the chessmen' (*PG*, p. 468). (More will be said about the 'inessential statements about the chessmen' in section 6.3.)

With remarks such as these, one is drawn towards the thorny issue of revisionism; and something must be said about Wittgenstein's idea that there should be no such thing as a foundational 'theory'. According to him, set theory (or, for that matter, the class theory of *Principia Mathematica*—Wittgenstein does not distinguish the two) is

[28] Another related topic that I shall not discuss is Gödel's incompleteness theorems (Gödel 1967). Wittgenstein's comments on these have been dismissed as containing errors, and displaying an inadequate understanding of the technical subtleties of these results. For a typical example, see Dummett (1978a: 166). There have been recent attempts to defend Wittgenstein. Stuart Shanker adopted the line that Wittgenstein was concerned not with Gödel's proofs directly (on this, see *RFM* vii, § 19) but with the accompanying philosophical remarks (Shanker 1988). Juliet Floyd goes further and defends Wittgenstein's understanding of Gödel's results by embedding his discussion in the larger discussion of impossibility proofs in mathematics, thus bringing into the picture Wittgenstein's remarks on geometry (in particular the remarks on the algebraic proof of the impossibility of the trisection of an angle) (Floyd 1995). I believe that this is on the right tracks, and would claim further, on the basis of remarks made by Kreisel—correcting his own earlier negative judgement in Kreisel (1958a: 153)—not only that Wittgenstein understood perfectly well Gödel's results, but also that some of his remarks could have turned out to have heuristic value. Kreisel's remarks are found in Kreisel (1950: 281 n.) and Kreisel and Takeuti (1974: 47–8); he also refers to them in Kreisel (1986: 107–8). (It is worth noticing that the first reference was published while Wittgenstein was still alive.)

a 'calculus' which is surrounded by 'clouds of thought'—the prose and its 'conceptual confusions' (*PI*, p. 232)—and philosophy is the activity of dissipating these 'clouds': 'What set theory has to lose is rather the atmosphere of clouds of thought surrounding the bare calculus' (*PG*, p. 470). (To call the system of *Principia Mathematica* (or set theory for that matter) a 'calculus' is to stretch the use of this term a bit, since logicists were interested precisely in, so to speak, 'algorithm-free' theories.) The result of philosophical activity is, therefore, not the rejection of the calculus on mathematical grounds; we should be left only with the bare calculus, deprived of the false conceptions embodied in the prose which gave it its presumed importance.

Therefore, although Wittgenstein claimed that the system of *Principia Mathematica* is, *qua* calculus, 'all right' (*WVC*, p. 114), he added that the motivation behind the construction of the calculus is 'wrong':

> Of course, when Russell was constructing his calculus he did not intend to develop merely a game of chess, but meant to reproduce with his calculus what the word 'infinite' really meant when it is applied. But in this he was wrong. (*WVC*, p. 114)

Wittgenstein was never interested in a 'mathematical' critique of *Principia Mathematica*. He wrote in the 1940s:

> It is my task, not to attack Russell's logic from within, but from without.
>
> That is to say: not to attack it mathematically—otherwise I should be doing mathematics—but its position, its office. (*RFM* vii, § 19)

But although Wittgenstein refused to attack a theory on mathematical grounds, he could at least shed doubt on its philosophical or foundational value. In his own jargon, if calculations cannot be wrong, at least their interest can be put in doubt: 'What I am doing is, not to shew that calculations are wrong, but to subject the *interest* of calculations to a test' (*RFM* ii, § 62). Clearly, Wittgenstein did not want to question the cogency of set theory: instead, he wished to clear up confusions surrounding it. In a perfectly explicit passage of the *Philosophical Grammar*, he starts by claiming that set theory is an attempt at 'describing' the actual infinite, as it cannot be grasped directly by a (naïve, as opposed to formalized) mathematical symbolism:

> Set theory attempts to grasp the infinite at a more general level than the investigation of the laws of the real numbers. It says that you can't grasp the

actual infinite by means of mathematical symbolism at all and therefore it can only be described and not represented. (*PG*, p. 368)

Since, as we saw, 'we can't describe mathematics, we can only do it' (*PR*, § 159), to conceive a calculus *as a description* is, according to Wittgenstein —who refers here to Frege's discussion of formalism in the *Grundgesetze der Arithmetik* (see section 6.3)—a complete misunderstanding of its nature: 'When set theory appeals to the human impossibility of a direct symbolisation of the infinite it brings in the crudest imaginable misinterpretation of its own calculus' (*PG*, p. 469). And, Wittgenstein goes on, such misinterpretation is, of course, at the origin of the calculus itself.[29] This does not prove that the calculus is incorrect but only that it is 'at worst uninteresting':

It is of course this very misinterpretation that is responsible for the invention of the calculus. But that doesn't show the calculus in itself to be something incorrect (it would be at worst uninteresting) and it is odd to believe that this part of mathematics is imperilled by any kind of philosophical (or mathematical) investigations. (*PG*, pp. 469–70)

So, the effect of philosophical activity will neither be the discovery of a 'technical' flaw nor the construction of a new calculus but simply clarity, and the result of clarity will be the abandonment of some foundational programmes and associated calculi such as logicism and the system of *Principia Mathematica*. The foregoing remarks explain Wittgenstein's annoyance at the news, in the mid-1920s, that Russell was planning a new edition of *Principia Mathematica*. In a letter to Ogden from Austria, Ramsey reported that Wittgenstein, with whom he was at that point having daily conversations, 'thought that he had shown R[ussell] that it was so wrong that a new edition was futile' (*LO*, p. 74). (Some of Wittgenstein's reasons will be explored in Chapters 2 and 3.) It also explains Wittgenstein's response to Hilbert's catchphrase about Cantorian set theory in his famous paper, 'On the Infinite'. In defiance of Brouwer's attacks against set theory (about which see sections 6.1–.2 and 7.2), Hilbert claimed: 'No one shall be able to drive us from the paradise that Cantor created for us' (Hilbert 1967b:

[29] Wittgenstein made a similar point in 1939: 'The misunderstandings we are going to deal with are misunderstandings without which the calculus would never have been invented, being of no other use, where the interest is centred entirely on the words which accompany the piece of mathematics you make' (*LFM*, p. 16).

376). To this, Wittgenstein replied repeatedly: 'I would try to show you that it is not a paradise—so that you'll leave of your own accord' (*LFM*, p. 103; *LA*, p. 28; *AWL*, p. 225).[30]

Thus Wittgenstein's position appears at first to be nothing more than plain common sense: it is not the job of the philosopher *qua* philosopher to intervene on the 'technical' side of a theory such as set theory since he has usually no qualifications to do so. But this is far from being a complete picture of Wittgenstein's argument. Indeed, there remain many things to be said about one non-negligible dimension of the theory: its prose. Such a dimension is not inessential: it is precisely the confusions originating in the prose that are responsible for the creation of the calculus itself, and if one eliminates these confusions, then there is no more need for the calculus, even if *qua* calculus it is correct. Between superficial, vacuous comments on mathematics on the one hand and full-blown 'revisionism' on the other (i.e. the rejection of parts of mathematics on what turn out ultimately to be purely philosophical grounds), there is thus room, according to Wittgenstein's conception, for a critical assessment of foundational programmes. This is the thought expressed in the very last paragraph of the *Philosophical Investigations*:

> An investigation is possible in connection with mathematics which is entirely analogous to our investigations of psychology. It is just as little a *mathematical* investigation as the other is a psychological one. It will *not* contain calculations, so it is not for example logistic. It might deserve the name of an investigation of the 'foundations of mathematics'. (*PI*, p. 232)

There is indeed no need for 'revisionism' if it is understood that foundational systems such as that set forth in *Principia Mathematica* (the main target of Wittgenstein's criticism) are not part of mathematics *stricto sensu*: a logical system embodying number theory is not number theory *per se* and *Principia Mathematica* is only an interpretation of Peano Arithmetic. At the time many thought, perhaps rightly so, that mathematical theories would survive any fatal criticisms levelled against their foundational counterparts. For example, one finds the idea already in Poincaré: when Russell stated that until 'the complete solution of our difficulties [the paradoxes] . . . is found we cannot be sure how much mathematics it will leave intact' (Russell 1906: 53),

[30] Whether Wittgenstein has provided arguments sound enough to convince anyone to leave Cantor's paradise is another matter. It is not my intention to try to evaluate Wittgenstein's claims at this stage.

Poincaré replied that 'only [Cantorian set theory] and [logicism] are called into question; real mathematics . . . will continue to develop according to their own principles' (Poincaré 1906: 307). Wittgenstein seemed to have been of the same opinion when he wrote about Russell's paradox: 'The Russellian contradiction is disquieting, not because it is a contradiction, but because the whole growth culminating in it is a cancerous growth, seeming to have grown out of the normal body aimlessly and senselessly' (*RFM* vii, § 11). It seems indeed that Wittgenstein believed that he could simply criticize *Principia Mathematica* (or set theory) without being forced to adopt an ultimately revisionist standpoint: the effect of philosophical clarity would be like the excision of a cancerous growth, leaving the body intact. It is a matter of controversy whether or not he erred in believing this, and I am aware that the foregoing remarks do not settle the matter. I can only claim here that I gave a faithful representation of Wittgenstein's views.

The result of a proper philosophical investigation of key concepts found in the prose will be, according to Wittgenstein, the clarification of some inherent confusions, the elimination of some misleading pictures associated with these concepts. There is nothing preventing us from believing that from this improved understanding new concepts might be introduced, leading to further mathematical (or logical) studies. But Wittgenstein thought that the search for conceptual clarity would lead to a slower growth of theories. This is the meaning of an important passage from the *Philosophical Grammar*:

> What will distinguish the mathematicians of the future from those of today will really be a greater sensitivity, and *that* will—as it were—prune mathematics; since people will then be more intent on absolute clarity than on the discovery of new games.
>
> Philosophical clarity will have the same effect on the growth of mathematics as sunlight has on the growth of potato shoots. (In a cellar they grow yards long.) (*PG*, p. 381)

So, although Wittgenstein believed that he could criticize foundational programmes on a conceptual level in such a way that mathematical theories themselves would remain free of any excisions, he also believed that such an activity would actually slow their growth.[31]

[31] Whether or not such a remark is borne out by the development of mathematics in this century is not a subject I wish to discuss.

I am aware that the foregoing remarks do not constitute an adequate treatment of the issues surrounding Wittgenstein's remarks on the nature of philosophical inquiry and revisionism. But it is hoped that they will indicate the form that an appropriate treatment should take.

2

Logicism without Classes

2.1. OPERATIONS AND ARITHMETIC

In proposition 6.01 of the *Tractatus Logico-Philosophicus*, Wittgenstein introduces the general form of an 'operation', which he writes $\Omega'(\bar{\eta})$, with a curious piece of symbolism:

$$[\bar{\xi}, N(\bar{\xi})]' (\bar{\eta}) \qquad (=[\bar{\eta}, \bar{\xi}, N(\bar{\xi})])$$

This symbolism will become clearer in a moment. For now, we can see first that on the left-hand side, the symbol Ω in $\Omega'(\bar{\eta})$ has been simply replaced by $[\bar{\xi}, N(\bar{\xi})]$, which presents in a more detailed manner the form of the operation. Secondly, we can see that the right hand simply presents an equivalent symbolization, where $(\bar{\eta})$ in $\Omega'(\bar{\eta})$ has been included within the square brackets of the left-hand symbol. It should now be pointed out that replacing $\bar{\eta}$ on the left-hand side by the set $\bar{P}$ of elementary propositions which appeared a little earlier in proposition 6 gives:

$$[\bar{\xi}, N(\bar{\xi})]' (\bar{P}) \qquad (=[\bar{P}, \bar{\xi}, N(\bar{\xi})])$$

We now recognize on the right-hand side the general form of the proposition which is presented in proposition 6:

$$[\bar{P}, \bar{\xi}, N(\bar{\xi})]$$

We thus see immediately that the general form of a proposition is a *particular case* of the general form of an 'operation'. The right-hand side representations of both the general form of an 'operation' and a proposition are obviously modelled on the expression

$$[a, x, 0'x]$$

found in 5.2522, where it is presented as a 'variable' representing the general term of a series which consists only of iterations of *0* with respect to *a*:

$$a, 0'a, 0'0'a, 0'0'0'a, 0'0'0'0'a, \ldots$$

So in the 'variable' *a* represents the first term of the series, *x* an arbitrary term of the series, and *O'x* the form of the term immediately following *x* in the series. From this and the fact that $\bar{\xi}$ is the notation for the set of all values of ξ (*TLP*, 5.501), we can infer that the general form of the proposition reads thus: any given (complex) proposition, from the set $\bar{\xi}$ of propositions is the result of successive applications of the operation N (about which I shall say more in a moment) to given propositions taken from the set of elementary propositions $\bar{P}$. In short: 'Propositions are truth-functions of elementary propositions' (*TLP*, 5). Wittgenstein also expressed the thought in these words: 'Every proposition is the result of truth-operations on elementary propositions' (*TLP*, 5.3); I shall come back to the notion of 'truth-operation'. Immediately after introducing the general form of an 'operation' in 6.01, Wittgenstein provides a definition of the natural numbers (integers) in 6.02. This definition is said in 6.021 to show that a number is an 'exponent of an operation' (*TLP*, 6.021). We also learn in the following remark, 6.022, that the concept of number is nothing more than the general form of number, which is presented immediately after, in 6.03:

$$[0, \xi, \xi + 1]$$

This 'form' is again modelled on the 'variable' given in 5.2522. (An explanation will be given shortly.) We thus see that the notion of an 'operation' also plays a crucial role in the definition of natural numbers.

So the notion of 'operation' plays a pivotal role in the symbolism of the *Tractatus Logico-Philosophicus*: on the one hand, truth-functions are based on 'truth-operations'; on the other, numbers are 'exponents of an operation'. Considering that 'operations' seem to be so central, it is amazing to notice how little is understood of Wittgenstein's remarks: not enough attention has been paid in the past to the curious piece of symbolism of 6.01. Well-known commentaries of the *Tractatus Logico-Philosophicus* by Elizabeth Anscombe and Max Black contained inadequate presentations (Anscombe 1971: 124–6; Black 1964: 313–14, 342–3).[1] The first adequate treatment of the definition of natural numbers in 6.02–3 and 6.241, in terms of what can be called a 'theory of operations', was given only very recently, in the first chapter of Lello Frascolla's

[1] For effective criticisms, see Frascolla (1994: 1–7).

Wittgenstein's Philosophy of Mathematics (Frascolla 1994). It seems plain that Wittgenstein intended his notion of 'operation' to play a fundamental role: to provide him with the basis from which he could introduce both the calculus of truth-functions in logic and a treatment of arithmetic which he saw as *not* being class-theoretical. In terms of the remarks made in Chapter 1, the notion of truth-function is 'algorithmic', since it was at least intended as a procedure of decision concerning logical truth. Indeed, the method of truth-tables was conceived by Wittgenstein as a 'mechanical expedient to facilitate the recognition of tautology, where it is complicated' (*TLP*, 6.1262); one must be able to decide mechanically *am Symbol allein* whether a proposition is tautological or not (*TLP*, 6.113, 6.126).

What Wittgenstein says about the notion of 'operation' very much resembles informal explanations of the notion of 'operator'. Two differences with the set-theoretic notion of function were mentioned in Chapter 1: first, an operator is defined by describing how it transforms an input without defining the set of inputs, that is without defining its domain. Secondly, there is no restriction on the domain of some operators. In particular, in lambda-calculus some operators are self-applicable, which is not the case with functions. Similar points are made by Wittgenstein, who insists that '[o]peration and function must not be confused with one another' (*TLP*, 5.25): to begin with, he defines an operation as 'the expression of a relation between the structures of its results and its bases' (*TLP*, 5.22),[2] i.e. that an operation is defined by showing how it takes an input and transforms it into an output (and thus obviously not by a set of ordered pairs . . .) Moreover, 'an operation does not assert anything'; only its result does (*TLP*, 5.25). (This is the origin of the notions of 'induction' and 'hypothesis' in the transitional period, which are introduced in Chapter 4 and discussed at length in Chapter 5.) Secondly, a function cannot be its own argument (this is the basis for Wittgenstein's alleged solution of Russell's paradox in 3.333) but 'the result of an operation can be its own basis' (*TLP*, 5.251; *WVC*, p. 217). The notion of 'operation' is intensional; it led Wittgenstein to

[2] Wittgenstein also points out that 'the concept of the operation is quite generally that according to which signs can be constructed according to a rule' (*NB*, p. 90), and that '[t]he operation is that which must happen to a proposition in order to make it into another. And that will naturally depend on their formal properties, on the internal similarity of their forms' (*TLP*, 5.23–5.231).

a definition of natural numbers which foreshadows the lambda-calculus definition given by Church and others more than ten years after the publication of the *Tractatus Logico-Philosophicus*.[3]

In Wittgenstein's terminology, a formal law is an 'operation' or an 'internal relation'. The most important feature of the notion of internal relation, which is introduced in *TLP*, 4.122, is that two entities are said to be internally related if it is inconceivable that they do not stand in such a relation: 'A property is internal if it is unthinkable that its objects do not possess it' (*TLP*, 4.123). Wittgenstein defines *Formenreihe* or 'formal series' (perhaps 'series of forms' is a better translation here) as series 'which are ordered by *internal* relations'. (He also defines the internal relation 'that orders a series' as 'equivalent to the operation by which one term arises from the other' (*TLP*, 5.232)!) The formal series or series of forms $a\mathrm{R}b$, for example,

$\exists x\ (a\mathrm{R}x\ \&\ x\mathrm{R}b)$
$\exists x, y\ (a\mathrm{R}x\ \&\ x\mathrm{R}y\ \&\ y\mathrm{R}b)$
$\exists x, y, z\ (a\mathrm{R}x\ \&\ x\mathrm{R}y\ \&\ y\mathrm{R}z\ \&\ z\mathrm{R}b)$
and so on

gets its form—is generated by—an internal relation (*TLP*, 4.1252, 4.1273). According to Wittgenstein, if *b* stands in such a relation to *a*, then '*b* is a successor of *a*'. His point is that it is only with an operation which can take its own result as its basis that 'progress from term to term in a formal series [is] possible' (*TLP*, 5.252). Thus one can progress in a series from a given point of departure. With *a* not being the result of an application of the operation *0'ξ* (the symbol *0* is here the variable for the relation; the comma, to which we shall return, represents the result of the application; and the symbol *ξ* represents the variable for any point of departure), the expression *0'a* represents the result of one application of the operation. The operation can be applied to this result, forming thus the expression *0'0'a*, and so on. (On the equivalence between the notion of operation and that of the 'and so on', see the end of section 2.3 below.) These repeated applications of the same operation Wittgenstein calls 'successive application'; thus *0'0'0'a* is the result of the 'threefold succes-

[3] George Boolos had made a remark to that effect in conversation many years earlier. Frascolla's presentation in ch. 1 of *Wittgenstein's Philosophy of Mathematics* makes the connection with lambda-calculus patent, but only mentions it in a footnote (Frascolla 1994: 176), perhaps because it appears so obvious to him.

sive application of "$0'\xi$" to "a"' (*TLP*, 5.2521). Thus successive applications of $0'\xi$ to a base a will form the series

$$a, 0'a, 0'0'a, 0'0'0'a, 0'0'0'0'a, \ldots$$

and, since '[a]n operation shows itself in a variable; it shows how we can proceed from one form of proposition to another' (*TLP*, 5.24), one may introduce the 'variable' $[a, x, 0'x]$ of 5.2522, which was presented earlier.

The truth-functions of elementary propositions are defined by Wittgenstein as 'results of operations which have the elementary propositions as bases' (*TLP*, 5.234). These particular operations he calls 'truth-operations' (*TLP*, 5.234), and they are: 'denial', 'logical addition', 'logical multiplication', etc. (*TLP*, 5.2341). It is at this point that Wittgenstein introduces one of the technical innovations of the *Tractatus Logico-Philosophicus*, the operation of joint negation, which is equivalent to the better-known Sheffer stroke (Sheffer 1913):[4]

> Every truth-function is the result of the successive application of the operation (- - - -T)(ξ, ...) to elementary propositions.
>
> This operation denies all the propositions in the righ-hand bracket and I call it the negation of these propositions. (*TLP*, 5.5)

In order to understand this quotation, it should be recalled that Wittgenstein rewrites the truth-table for p & q:

p	q	p & q
T	T	T
T	F	F
F	T	F
F	F	F

by transforming the last column in a row: (TFFF)(p, q) (*TLP*, 5.101). For ξ, the notation for the values of the variable ξ, Wittgenstein writes $N(\bar{\xi})$ for their joint negation (*TLP*, 5.502), hence the name 'N operator'. So, if there is only one value of ξ, say, p, then $N(\bar{\xi}) = \neg p$; if there are two, say, p and q, then $N(\bar{\xi}) = \neg p \, \& \neg q$, and so on (*TLP*, 5.51); so the N operator would be written (FFFT)(p, q). (I shall come back to it in the next section.) There is now no mystery

[4] Amazingly enough, in his introduction to the 2nd edn. of *Principia Mathematica*, Russell presented the Sheffer stroke as the 'most definite improvement resulting from work in mathematical logic' since the first edition had appeared (*PM*, p. xiii)!

left behind the symbolic representation of the general form of the proposition $[\bar{P}, \bar{\xi}, N(\bar{\xi})]$. There are, however, some difficulties linked with the fact that this notation implies, by virtue of being a variant of the 'variable' $[a, x, 0'x]$ of 5.2522, that the propositions are ordered in a formal series. This cannot be the case, since, for example, *p* & *q* has more than one immediate predecessor (either *p* or *q*) so there cannot be a total ordering. At any rate, it suffices for our purposes that the fundamental role played here by the notion of operation be made patent. I shall now turn to natural numbers and arithmetic (*TLP* 6.2–3, 6.241), following closely Frascolla's explanations in *Wittgenstein's Philosophy of Mathematics* (Frascolla 1994: 8–23). His conjecture is that Wittgenstein sets up in these propositions a 'theory of operations', with a view to showing that a given equation

$$t = s$$

is a theorem of arithmetic if and only if one has a proof of the corresponding equation

$$\Omega^{t}x = \Omega^{s}{}'x$$

in the theory of operations (Frascolla 1994: 3). In other words, Wittgenstein had in mind, *avant la lettre*, a *reduction* of arithmetic to his 'theory of operations'. (Of course, there is no question that Wittgenstein could only and would probably only want to show how to proceed—i.e. in the style of the book, give rough indications.) To talk of 'theories' and 'reductions' is in a way an 'unwittgenstinian' *façon de parler*, but apologies need not be made because this conjecture makes much sense if we think of it the following way. If Wittgenstein could show that natural numbers and elementary arithmetic operations can be defined in terms of operations, then he would have an argument against Russell, who (so to speak) attempted a reduction to the theory of classes ('translation' would be a more appropriate term here). It is no surprise that the brief but powerful train of thought which starts in 6.02 with the definition of natural numbers (actually integers) and finishes with the general form of number $(0, \xi, \xi + 1)$ in 6.03 is actually followed immediately by the famous remark that 'the theory of classes is altogether superfluous in mathematics' (*TLP*, 6.031): Wittgenstein can make such a claim only because he has shown in 6.02–3 that one can do without classes. But one still senses here the remnant of a 'reductionist' attitude, hence the idea of speaking of a 'logicism without classes'.

The first part of proposition 6.02 provides the following obscurely put inductive definition:

$$x =_{\text{def.}} \Omega^{0\text{'}}x$$
$$\Omega\text{'}\Omega^{\nu\text{'}}x =_{\text{def.}} \Omega^{\nu+1\text{'}}x$$

In this definition the symbol *x* is a variable for any basis, i.e. for any term which is not generated by an application of a given operation. The symbol Ω is the variable standing for any operation, such as N above. It is accompanied by the single inverted comma, which is borrowed from the notation for descriptive functions in *Principia Mathematica*, where

$$\text{R}\text{`}y = (\iota x)\ (x\text{R}y)$$

reads 'the only term *x* which has the relation R to *y*'; so, supposing that R is the relation of father to son, R'*y* would mean 'the father of *y*' (*30.01). In Wittgenstein's case, the single inverted comma means something rather different: it is part, with Ω (or earlier with *0* in *0'0'0'0'a*, for example) of the notation representing the form of the result of an application of the operation for which Ω is the variable. More difficult is the interpretation of the symbols 0 and ν + 1. As Frascolla points out, we must assume that terms of the form 0 + 1 + 1 + . . . + 1 and thus 0 and ν + 1 are available within the language of the theory of operations. Here 0 is the symbol for the form with no occurrences of + 1, and ν the symbol for the form with *n* times the occurrence of + 1. Wittgenstein's choice of sign is unfortunate as it causes confusion, especially since + is also used as the symbol for the operation of addition in arithmetic. To avoid confusion, I here follow Frascolla in introducing the terms *0*, *S0*, *SS0*, *SSS0*, . . . (these terms are in italics in order to distinguish them from 0, S0, SS0, SSS0, . . ., which are the numerals in arithmetics), and replacing *x* with $\Omega^{0\text{'}}x$, Ω'*x* with $\Omega^{S0\text{'}}x$, Ω'Ω'*x* with $\Omega^{SS0\text{'}}x$, etc., so that the series

$$x,\ \Omega\text{'}x,\ \Omega\text{'}\Omega\text{'}x,\ \Omega\text{'}\Omega\text{'}\Omega\text{'}x,\ \ldots$$

is to be understood as being composed of *x*, which is the form of any expression which is the initial expression of a series generated by successive applications of an operation (thus not itself generated by an application of the operation); Ω'*x*, which is the form of any expression which is the result of a single application of an operation to an initial expression; Ω'Ω'*x*, which is the form of any expression which is the result of a twofold successive application of an operation, i.e.

an application of the operation to the result of its own application to an initial expression, and so on . . . Since Ω^{0}'x is identified with x, one can say that *0* stands for the formal property constituted by the number of times (in this case, zero) the operation is applied to produce such an expression, and it provides the meaning of the numeral 0 in arithmetic. Similarly, Ω^{S0}'x is identified with Ω'x, and one can say that *S0* stands for the formal property constituted by the number of times the operation is applied to produce such an expression (in this case a single application to the base) and it provides the meaning of the numeral S0 in arithmetic. Again, the term Ω^{SS0}'x is identified with Ω'Ω'x, and one can say that *SS0* stands for the formal property constituted by the number of times the operation is applied to produce such an expression (in this case a twofold successive application to the base) and it provides the meaning of the numeral SS0 in arithmetic. So Wittgenstein, by introducing the corresponding terms as exponents of Ω, reduces arithmetical numerals to the notion of the successive application of an operation, thus justifying his calling numbers 'exponent of an operation' (*TLP*, 6.021). The inductive definition thus succeeds in reducing the notions of zero and successor to the notion of an application of an operation. This is broadly similar to the introduction of natural numbers in lambda-calculus as, say, the iterators, *n* being represented by:

$$\lambda \mathrm{f}.\lambda x\ \underbrace{\mathrm{f}(\mathrm{f}\ \ldots\ \mathrm{f}(}_{n}\mathrm{f}(x)\ \ldots\ )$$

(For details, see Barendregt (1984: § 6.2.9).)

The second half of 6.02 presents an introduction of numbers 1, 2, and 3 (using his expressions '*0, S0, SS0, . . .*'), and Wittgenstein is then only one step away from the general concept of number: 'the concept of number is simply what is common to all numbers, the general form of a number' (*TLP*, 6.022). In the series formed by the expressions *0*, *S0*, *SS0*, . . . the relation of each term to its immediate successor is uniform; it is, as Wittgenstein would say, 'what is common to all'. This uniform aspect can be expressed in turn in accordance with the explanations of construction of the 'variable' [*a*, *x*, *0'x*] in 5.2522, which applies whenever a formal concept constitutes a series; and, since Wittgenstein used terms of the form $0 + 1 + 1 + \ldots + 1$ as exponents of the operation, the general form of number he gives is $[0, \xi, \xi + 1]$. Having obtained his general form, Wittgenstein has completed his reduction of natural numbers to the notion of successive applications of an operation. As I mentioned earlier, he is thus free to claim in the very next sentence that 'the theory of classes is altogether superfluous in mathematics' (*TLP*, 6.031). This conclusion is further supported by the possibility of a reduction of the arithmetical operations of addition and multiplication to the notion of successive application of an operation, following indications given in his 'proof' of the arithmetical equation $2 \times 2 = 4$ in proposition 6.241.[5] One may add here that Wittgenstein's intensional notion of number is at the heart of his logicism without classes. To put it in words less likely to mislead: it is at the heart of his anti-logicist standpoint.

2.2. TRUTH-FUNCTIONS, GENERALITY, AND INFINITY

One of the best-known theses of the *Tractatus Logico-Philosophicus* is that '[t]he proposition is a picture' or a 'model of reality' (*TLP*, 4.01). Contrary to Max Black's claim that the picture theory applies primarily to elementary propositions (Black 1964: 220), Merrill and Jaakko Hintikka have argued convincingly in *Investigating Wittgenstein* that it also applies to complex propositions (Hintikka and Hintikka 1986: chs. 4–5).[6] Wittgenstein's strategy is easily understood: since proposition 6 tells us that the general form of the proposition is $[\bar{P}, \xi, N(\xi)]$, i.e. that any proposition is the result of successive applications of the operation $N(\xi)$ to the elementary propositions $\bar{P}$, the task of extending pictoriality to complex propositions is reduced to that of showing that a conjunction of pictures remains a picture and that the negation of a picture also remains a picture (negation and conjunction being the two truth-functions to which the others are reduced with the help of the N operator). As far as conjunction is concerned, it is obvious that two pictures can be put together to form a larger one. As for negation, the thesis of the bipolarity of propositions helps us to see that the negation of a picture is not only a picture, it is the same picture, with, so to speak, converse polarity ('[d]enial reverses the sense of a proposition', *TLP*, 5.2341).

[5] A correct reconstruction of 6.241 is given in Frascolla (1994: 13–19). The 'proof' is, with one exception, perfectly in order, and Black's judgement that it is 'eccentric and would not satisfy contemporary standards of mathematical rigour' (1964: 343) just betrays his lack of understanding of it.

[6] See also Hintikka (1994: 241–3).

To extend the picture theory, Wittgenstein must also deal with quantified and identity statements. Only then would he be in a position to claim that pictoriality is the 'essence' of language. As I take it, his strategy was to reduce quantifiers to conjunctions and disjunctions and to eliminate the identity sign altogether. I shall look in turn at these two moves, which are full of consequences for the philosophy of mathematics. Quantification and the associated topics of validity and infinity will be dealt with in this section, while identity will be looked at within the larger framework of debates around arbitrary functions in section 3.1.

In the 1920s, a fundamental debate on the nature of quantifiers took place, fuelled by the *Grundlagenstreit* between Hilbertian formalists and Brouwerian intuitionists. (This quarrel will be discussed in section 4.1.) As we shall see in section 2.3, one of the major players in that debate was Frank Ramsey, who found Wittgenstein's treatment of quantification so sound that he repeatedly made it his own (Ramsey 1978: 28–9, 54–7, 159–60). (Ramsey made crucial use of Wittgenstein's infinite conjunctions and disjunctions in his attempt to renovate Russell's theory of types.) But very few scholars paid attention to the quantifiers in the *Tractatus Logico-Philosophicus*.[7] The universal quantifier as we know it was introduced by Frege in his *Begriffsschrift*, in 1879.[8] According to him, the universal quantifier is a second-level concept, with one argument place to be filled up by any first-level predicate. The result of filling the place with any such predicate would be a sentence which would be determinately true or false. The truth-value would then be a product of truth-values, which is the result of applying the predicate to each object of the domain of the variable bound by the quantifier. If the result of applying the predicate to all the objects is always true, then the quantified statement is true; but if the result is false only once, then it is false. When the domain of quantification is finite (and surveyable) then the truth-value of the quantified statement could in principle be determined as a finite product, but when the domain of quantification is infinite (or finite but unsurveyable) this assumption would be open to question, because it would be claimed that we cannot determine its

[7] Warren Goldfarb, in his excellent study 'Logic in the Twenties: The Nature of the Quantifier', mentions the *TLP* only once (Goldfarb 1979: 353). A notable exception is Anscombe (1971, ch. 11).

[8] See Frege (1967: §§ 11–12). For a sketch of Frege's conception of quantification, see Dummett (1981, chs. 2, 15).

truth conclusively. This was particularly annoying for Frege, since for him the domain of quantification was always the infinite totality of objects: a quantified statement would always be an infinite product of truth-values. Frege circumvented this difficulty by replying that we only need to have a general grasp of the totality—in Dummett's words, that 'we, as it were, survey it in thought as a whole' (Dummett 1981: 517).[9] This is the reason why Frege would say of a generalized statement such as 'All men are mortal' that the fact that it implies that Chief Akpanya, who is unknown to him, is mortal is not part of the thought asserted by this proposition.[10] Frege insisted that it is sufficient that we have a grasp of the meaning of the quantified statement, of the 'thought' expressed by it. If it were possible to speak at all of a semantic theory here, it would be correct to say that at the semantic level it is implied by Frege's account that the universal quantifier amounts to a logical product, and the existential quantifier to a logical sum; but he did not make this connection since for him it is not the case at the level of meaning (sense).[11]

Russell adopted the Fregean account of quantifiers but, following Peano, introduced a new symbolism for them. In *Principia Mathematica*, for any propositional function $\phi\hat{x}$ there is a range of values which consists of all the propositions which can be obtained by giving every possible determination of x in ϕx. A value of x 'satisfies' $\phi\hat{x}$ if it renders ϕx true. The fact that every possible determination of x renders ϕx true is symbolized by $(x).\phi x$, which may be read as 'ϕx always' or 'ϕx is always true'. The fact that some determination of x renders ϕx true is symbolized by $(\exists x).\phi x$, which may be read as 'there exists an x for which ϕx is true' or 'there exists an x satisfying $\phi\hat{x}$'. These new symbols are not defined, they embody new primitive ideas (*PM* i, p. 15). (These symbols correspond respectively to $\forall x\ \phi(x)$ and $\exists x\ \phi(x)$, which are used throughout the book.) Although Russell did not make Frege's distinction between sense and reference, he explained quantification in a similar way. Indeed, he considered propositions such as 'All men are mortal' as a particular kind of judgement, which he called 'general judgements', and of these he said:

[9] This is the reason why Frege's account cannot properly be said to be similar to the substitutional interpretation.

[10] This remark was made in the review of Edmund Husserl's *Philosophie der Arithmetik* (Frege 1984: 205). See also Frege (1980a: § 47).

[11] For the distinction between the semantic and meaning levels, see Dummett (1981: 57).

> Our judgement that all men are mortal collects together a number of elementary judgements. It is not, however, composed of these, since (e.g.) the fact that Socrates is mortal is no part of what we assert, as may be seen by considering the fact that our assertion can be understood by a person who has never heard of Socrates. In order to understand the judgement 'all men are mortal,' it is not necessary to know what men there are. (*PM* i, p. 45)

The idea expressed here is similar to that of Frege inasmuch as they both claim that there is no need to have a complete knowledge of the domain (range of value) in order to understand the generalized proposition or judgement.

In the *Tractatus Logico-Philosophicus* Wittgenstein developed both a theory of quantification and a theory of generality. He initially approved the use of the Russellian symbolism $(x).\phi x$ or $\forall x\ \phi(x)$ for expressing generality, because he claimed that other symbolisms run into difficulties (*TLP*, 4.0411).[12] But his own treatment is rather different from that of Russell, who pointed out this very difference in his introduction:

> Wittgenstein's method of dealing with general propositions [i.e. '$(x).fx$' and '$(\exists x).fx$'] differs from previous methods by the fact that the generality comes only in specifying the set of propositions concerned, and when this has been done the building up of truth-functions proceeds exactly as it would in the case of a finite number of enumerated arguments $p, q, r \ldots$ (*TLP*, p. 14)

We saw that the general form of the proposition is $[\bar{P}, \bar{\xi}, N(\bar{\xi})]$, and that by this Wittgenstein meant that any proposition is the result of successive applications of the operation $N(\bar{\xi})$ to the elementary propositions of $\bar{P}$. For $\bar{\xi}$ the values of the variable ξ must be determined (*TLP*, 3.316, 3.317, 5.501) and the determination of the variables is the description of the propositions for which the variable stands. There are, according to 5.501, three possible kinds of description of these propositions: they can be given either (1) by a direct enumeration or (2) as values of a Russellian propositional function $F(x)$, i.e. by 'giving a function fx whose *values* for all values of x are the propositions to be described' (*TLP* 5.501) or (3) through a formal law which constructs the series of propositions. One important fact about these three kinds of description is that their differences were deemed unessential by Wittgenstein: 'How the description of the terms of the expression in brackets takes place is unessential' (*TLP* 5.501). For general propositions, where the values of the variable ξ are given as

[12] These difficulties are discussed in Anscombe (1971: 140–1).

the values of a function—as in *TLP* 5.52—say for any F(x) whose values are F(a), F(b) and F(c), N(F(x)) would be a proposition of the form:

$$\neg F(a) \wedge \neg F(b) \wedge \neg F(c)$$

This reading of ξ opens the door to a reading of the quantifiers not entirely without analogy to that already existing in the the algebraic tradition originating in the work of C. S. Peirce,[13] i.e. to a reading of $\forall x$ F(x) as a logical product and $\exists x$ F(x) as a logical sum:

$$\forall x\, F(x) \leftrightarrow F(a) \wedge F(b) \wedge F(c) \wedge \ldots$$
$$\exists x\, F(x) \leftrightarrow F(a) \vee F(b) \vee F(c) \vee \ldots$$

Wittgenstein said as much in his 1932–3 lectures, according to G. E. Moore's notes: 'He said that there was a temptation, to which he had yielded in the *Tractatus*, to say that (x).fx is identical with the logical product "fa . fb . fc . . .", and (∃x).fx identical with the logical sum "fa ∨ fb ∨ fc . . .", but that this was in both cases a mistake' (*M*, p. 297). But in reply to a letter from Russell dated 13 August 1919, Wittgenstein seems to be making a different claim: 'You are quite right in saying that "N($\bar{\xi}$)" may also be made to mean $\neg p \vee \neg q \vee \neg r \vee \ldots$ But this does not matter! I suppose you don't understand the notation "ξ". It does not mean "for all values of ξ . . ." ' (*NB*, p. 131). As I see it, the reasons for Wittgenstein's claim that 'ξ' does not mean 'for all values of ξ . . .' have to do with the intensional outlook of the *Tractatus Logico-Philosophicus*. As we saw in the previous section, Wittgenstein was toying with a purely intensional notion of operation, and his definition of integers is in keeping with the spirit of the lambda-calculus. Like the founders of lambda-calculus, he simply did not think in terms of extensions, although these can be seen as equivalent (and there is no reason to believe that he would have denied this).

[13] Charles Sanders Peirce introduced in 1885, thus a few years after Frege, the universal and the existential quantifier as, respectively, a product and a sum: 'Here, in order to render the notation as iconical as possible we may use Σ for *some*, suggesting a sum, and Π for *all*, suggesting a product. Thus $\Sigma_i x_i$ means that x is true of some one of the individuals denoted by i or $\Sigma_i x_i = x_i + x_j + x_k$ + etc. In the same way, $\Pi_i x_i$ means that x is true of all these individuals, or $\Pi_i x_i = x_i x_j x_k$, etc.' (Peirce 1933: 228). It is interesting that Peirce used Σ and Π, which were already in systematic use in mathematics as variable binding operators from at least the early nineteenth century, and Cauchy's introduction of the idea of a limit. For example, Σ is used in the definition of an integral as an infinite sum of all values of a function.

By the same token, there is no reason to believe that Wittgenstein had any qualms with quantification over an infinite domain. Are there such domains in the *Tractatus Logico-Philosophicus*? In other words, are there infinite totalities in that work? In a hitherto unpublished manuscript, Michael Wrigley (1991) has argued convincingly for a positive answer, and I shall summarize his remarks here.[14] To begin with, Wittgenstein entertains in 4.2211 the possibility that there could be infinitely many objects: 'Even if the world is infinitely complex, so that every fact consists of an infinite number of atomic facts and every atomic fact is composed of an infinite number of objects, even then there must be objects and atomic facts' (*TLP*, 4.2211). Moreover, commenting on Russell's Axiom of Infinity (about which more will be said in section 3.2), he wrote that what it intended to say would show itself in language through the existence of infinitely many names with different meanings (*TLP*, 5.535). In virtue of the name–object relations in the *Tractatus Logico-Philosophicus*, an infinity of names implies an infinity of objects. Now, since objects form the substance of the world (*TLP*, 2.021) and they are unalterable (*TLP*, 2.023, 2.0271), they must be given as a determinate and unchanging totality, an infinite one if we follow the suggestion in 4.2211. This opens the door to actual infinity. Wittgenstein also wrote of the logical space, which is the space of all possible propositions, that it is an 'infinite whole' (*TLP*, 4.463). Since 'the whole logical space must be given' with a particular proposition (*TLP*, 3.42), it is implicit in Wittgenstein's doctrines that the totality of propositions be infinite and determinate, that is in a sense actual.

Two brief remarks are in order here. First, these remarks about infinity in the *Tractatus Logico-Philosophicus* cohere quite well with a remark made by Wittgenstein to Georg Kreisel, that he had put down his system in a finite setting without bothering about the infinite case, assuming that if a problem was to be found in the infinite case, then there would have already been a problem in the finite case. Propositions such as 'There are $\aleph_0$ objects' are excluded from the system of the *Tractatus Logico-Philosophicus* because 'object' is here a proper concept word, and 'it is senseless to speak of *the number of all objects*' (*TLP*, 4.1272). Being presumably faithful to his own *Diktat*, Wittgenstein simply did not think hard about this question. Secondly, Wittgenstein came to realize that this was one of the cru-

[14] See also Wrigley (1987).

cial mistakes in his book. To his student Desmond Lee he presented (some time in 1930–1) the idea of 'treating infinity as a number' (which is, as we shall see in section 6.3, an idea that he associates with the concept of actual infinity) and 'supposing that there can be an infinite number of propositions' as one source of his erroneous ideas about analysis in the *Tractatus Logico-Philosophicus* (*LWL*, p. 119) (the other source being a mistake about propositions expressing degrees of quality, which will be discussed at length in 5.1). To use a piece of Wittgenstein's own jargon which will be explained in a moment, his treatment of generality meant that even in cases where no enumeration is possible, as in (1) above (an infinite domain would be such a case), it is still possible to consider the universal quantifier as a logical product. As we shall see in section 4.2, Wittgenstein recognized upon his return to philosophy in 1929 that this was a grave mistake.

Another important aspect of Wittgenstein's treatment of quantification is that 'Wittgenstein links the essence of generality with the notion of a variable, rather than with that of a quantifier' (Black 1964: 282). Indeed, generality is distinguished in the *Tractatus Logico-Philosophicus* from truth-functions:

> I separate the concept *all* from the truth-function.
>
> Frege and Russell have introduced generality in connexion with the logical product or the logical sum. Then it would be difficult to understand the propositions '(∃x).fx' and '(x).fx' in which both ideas lie concealed. (*TLP*, 5.521)

Generality comes with the specification of the arguments as the values of a given propositional function, and differs from truth-functions *per se* which are the logical product for the universal quantifier and the logical sum for the existential quantifier. This distinction has to do with the notion of 'logical prototype' (*TLP* 5.522). In proposition 3.315, Wittgenstein tells us that an expression such as *a*R*b*, where all components would be changed into variables, is to be called a 'logical prototype'. Still following 3.315, one can also say that an expression such as *x*R*b*, *x* being a variable, collects all the propositions of the form —R*b*. This is why Wittgenstein said that the symbolism of generality 'makes constants prominent' (*TLP* 5.522), as Ramsey clearly saw (1978: 28–9). So given *a*R*b* we can pass to *x*R*b* and, once we have this as the set of all propositions of the form — R*b*, the expression ∀*x* (*x*R*b*) is just the proposition which is a truth-function of these, while generality is expressed by the use of the variable x. (This aspect of the *Tractatus Logico-Philosophicus* already shows the similarity of Wittgenstein's ideas to Skolem's primitive recursive arithmetic. This connection will be developed further in section 4.2.)

In 6.1232, Wittgenstein distinguished an 'essential' form of general validity from the 'accidental' form: 'Logical general validity, we could call essential as opposed to accidental general validity, e.g. of the proposition "all men are mortal" ' (*TLP* 6.1232). As we saw, logical propositions (tautologies) were thought by Wittgenstein to be all inductively generated from the elementary propositions by the N operator; therefore, although Wittgenstein thought that the fact that logical propositions are generally valid does not demarcate them from non-logical propositions (since to be general in logic means no more than to be 'accidentally valid for all things', *TLP*, 6.1231), logical propositions do in fact possess logical general validity. This 'essential' or 'logical' validity was also considered by Wittgenstein to be an essential feature of mathematical pseudo-propositions: it derives from the fundamental fact that members of a formal series are all generated by a formal law. But the theory of classes as he knew it from *Principia Mathematica* needs axioms such as the Axiom of Reducibility or the Axiom of Infinity which are admittedly not logical propositions. At best, these would possess only 'accidental' general validity, and this is the reason why Wittgenstein thought that class- or set-theoretical axiomatizations are 'superfluous' in mathematics. Proposition 6.031, when quoted fully, reads indeed:

> The theory of classes is altogether superfluous in mathematics.
>
> This is connected with the fact that the generality which we need in mathematics is not the *accidental* one. (*TLP*, 6.031)

In Wittgenstein's eyes Russell was obviously guilty of promoting confusion between 'accidental' generality (in the empirical case) and 'essential' generality (in the case of a series given by a formal law) by representing both by the same symbolism, that of the propositional function.[15] This reading is confirmed by the manuscript of a lecture by Waismann, where he expounded the philosophy of mathematics

[15] Wittgenstein was later to make the same point, saying that '(x) fx' was originally meant to symbolize statements of ordinary language such as 'All men are mortal', and was then extended to mathematical statements, 'where very different grammars apply' (*AWL*, p. 68).

of the *Tractatus Logico-Philosophicus* (Waismann 1986).[16] A famous symposium on the foundations of mathematics organized by the *Wiener Kreis* took place in Königsberg in 1931, where Rudolf Carnap presented the case for logicism, Arend Heyting the case for intuitionism and Johann von Neumann the case for Hilbertian formalism. Their papers, which have since acquired the status of classics,[17] were published in an issue of *Erkenntnis* along with a transcription of the conversations (where, incidentally, Kurt Gödel made the first public mention of his incompleteness results). It is barely known, however, that the three musketeers were four, since Waismann also presented in this symposium a paper entitled 'Über das Wesen der Mathematik; Der Standpunkt Wittgensteins', which he never submitted for publication (Waismann 1986). In the heavily annotated manuscript left by Waismann one finds a distinction between 'totality' and 'system' (which follows the lines set out in *TLP*, 5.25, 5.251), it is claimed that there is an essential difference between, say, the chairs in the conference room and natural numbers, which is correctly rendered by distinguishing between a 'totality', which is empirical and given by a propositional function, and a 'system', given by an operation: 'An empirical totality goes back to a *property* (a propositional function); a system to an *operation*' (Waismann 1986: 64). Wittgenstein's claim that generality in mathematics is not 'accidental' is echoed thus by Waismann: 'In mathematics we are always confronted with systems, and not with totalities' (p. 65). Wittgenstein's standpoint has been neglected. There are reasons for this neglect: first, it was never really developed by either Wittgenstein or Waismann (or anyone else for that matter), and, secondly, as a result no one really understood how it differed from other standpoints (the confusion with logicism was frequent). Thirdly, no one at the time was able to see the affinities between Wittgenstein's ideas and Church's lambda-calculus, which appeared in 1932.[18]

[16] Waismann's notes of the discussions with Schlick and Wittgenstein show that he had preparatory discussions with Wittgenstein in Vienna. See *WVC*, pp. 102–7, 213–17, which contains material quite similar to that of the manuscript of the Königsberg lecture. See also Granger (1990).

[17] Translations of these papers appear in Benacerraf and Putnam (1983: 41–65).

[18] The concept of 'operator' in lambda-calculus was itself never successfully formalized, in the sense that theories strong enough to serve any foundational purpose, such as Church's original one (Church 1932), turn out to be inconsistent, while weaker versions are lacking in comprehensiveness. So the concept of 'operator' is apparently not serviceable for a 'foundational' enterprise in the traditional sense. On the other

2.3. PREDICATIVITY

In the third part of the *Begriffsschrift*, Frege had already questioned Kant's thesis that all mathematical judgements are synthetic a priori.[19] But he remained hesitant at the time of the publication of the *Foundations of Arithmetic*, saying that he only hoped to 'have made it probable that the laws of arithmetic are analytic judgements and consequently a priori' (Frege 1980a: § 87). Russell and his French epigone Couturat were quick, however, to proclaim the Kantian doctrine dead. But neo-Kantianism was the dominant school on the Continent at the turn of the century and they found many opponents including illustrious mathematicians such as Henri Poincaré, who held steadfastly throughout his career to the Kantian idea that the principle of mathematical induction is synthetic a priori.[20] He considered suspicious the set-theoretical or class-theoretical approaches to natural numbers, and he notoriously objected that they could only issue in definitions that are viciously circular, for reasons which we will examine in a moment.

Russell had originally defined the notions of 'predicative' and 'non-predicative' in the following way: 'Norms (containing one variable) which do not define classes I propose to call *non-predicative*; those which do define classes I shall call *predicative*' (Russell 1906: 34); and Poincaré wanted to call 'non-predicative' all definitions implying a vicious circle (Poincaré 1906: 307).[21] Although Poincaré developed his criticisms partly on the basis of his explanation of Richard's Paradox,

hand, there is no need to insist on the importance of this concept for constructive mathematics, category theory, and computer science.

[19] In *The Semantic Tradition from Kant to Carnap*, Alberto Coffa (1991: pt. 1) has proposed that we look at the work of logicists such as Frege and Russell as part of a 'semantic tradition' originating in the writings of Bernard Bolzano, who was reacting against Kant's notion of 'pure intuition' (see also Coffa 1982). According to Kant, 'pure intuition' underlies synthetic a priori judgements of mathematics, but it was seen by followers of Bolzano as, roughly speaking, a semantically incoherent as well as unexplanatory notion. If all fundamental mathematical concepts could be defined in logical terms only, then it would appear that mathematical sentences, derived from these definitions, would be analytical in nature and not synthetic a priori. The notion of 'pure intuition' would thus be eliminated. So for authors within the 'semantic tradition' such as Frege, the strategy was, as far as arithmetic is concerned, to define natural numbers and the principle of mathematical induction in purely logical terms.

[20] See Poincaré (1952a: 12–13; 1906: 298, 313).

[21] Commentators have noticed that Poincaré gave, on different occasions, more than one definition of predicativity. See Heinzmann (1985: 31, 70).

it must be said that, contrary to Russell, he did not think that the circularity involved in the logicist definitions of the natural numbers is 'vicious' because it leads to paradoxes. Poincaré's charge of circularity was meant not so much to point to a logical defect of the definitions as to cast doubt on the fact that they are genuine definitions: 'A definition which contains a vicious circle defines nothing' (Poincaré 1952b: 191).[22] This is the basis for Poincaré's restricted form of 'definabilism', which is usually referred to as 'predicativism'. In one of his last papers, 'La logique de l'infini', Poincaré gave the following description of what it means for a definition to imply a 'vicious circle':

> These are again these definitions by postulates, but the postulate is here a relation between the object to be defined and *all* the individuals of a genus of which the object to be defined is itself supposed to be a member (or of which one supposes to be members objects which themselves can be defined only by the object to be defined) . . . According to pragmatists such a definition implies a vicious circle. (Poincaré 1963: 70)[23]

Although Poincaré introduced the idea of a 'vicious circle' while discussing Hilbert's first formulation of his metamathematical programme and its concomitant need for a consistency proof,[24] his charge of circularity was also directed at the multiple logicist

[22] Such definitions are frequent in classical mathematics. Informally a set S is defined impredicatively if given by $S = \{ x \mid \forall y \in A\ P(x,y) \}$, and if S is already in A. Notions as elementary as e.g. the least upper bound are defined impredicatively. See Weyl (1987: 111–12).

[23] When he wrote 'La Logique de l'infini', Poincaré described himself as a 'pragmatist', i.e. a practicioner of mathematics, as opposed to someone wishing to provide set-theoretic foundations. There is thus no reference to the philosophical school of 'pragmatism', and a more appropriate term would have been 'finitist', since Poincaré defines 'pragmatists' as mathematicians who 'wish to consider only objects which can be defined in a finite number of words' (Poincaré 1963: 66). This is the stance taken by most French mathematicians of the time. See the beginning of section 3.1 for details.

[24] For Hilbert's original programme, see Hilbert (1967a). Briefly, Poincaré's argument was to the effect that some form of induction was to be used inside metamathematics itself, in order to prove the validity of mathematical induction in ordinary, naïve arithmetic, and that this new form of induction had to be at least as strong as the ordinary form, therefore invalidating the whole process. It has been argued that Poincaré was vindicated by Gödel's second incompleteness theorem: the second form of induction could not be weaker than the first, but must be stronger. (See e.g. Jean van Heijenoort's introductory note to Weyl 1967.) It is worth noticing, however, that in Gentzen's proof of consistency of Peano Arithmetic a single application of a stronger form of induction, up to ε_0, applied to only one primitive recursive predicate, validates all the instances of ω-induction applied to predicates of any arithmetical complexity. This last point, put to me by Daniel Isaacson, is also discussed in Goldfarb (1988: 64–5).

attempts at a purely logical definition of mathematical induction. In the same paper he criticized Whitehead and Russell's proof of mathematical induction as it was formulated in an early paper of Whitehead (Whitehead 1902: § iii). Poincaré's argument was that this definition of an 'inductive number' was impredicative because it was defined with the help of the notion of 'inductive class' which already includes the notion of 'inductive number' (Poincaré 1952b: 190–1). He pointed out that such definitions could only be sustained if one adheres to the actual infinite:[25]

> It is the belief in the existence of actual infinity that has given birth to these non-predicative definitions. I must explain myself. In these definitions appears the word *all* . . . The word *all* has a very precise meaning when it is a question of a finite number of objects; but for it still to have a precise meaning when the number of objects is infinite, it is necessary that there should exist an actual infinity. Otherwise *all* these objects cannot be conceived as existing prior to their definition and then, if the definition of a notion N depends on *all* the objects A, it may be tainted with the vicious circle, if among the objects A there is one that cannot be defined without bringing in the notion N itself. (Poincaré 1952b: 194)

For Poincaré, that existence of an actual infinity was needed to make sense of such definitions was precisely the reason why he believed that they were viciously circular.[26]

Poincaré does not seem to have been aware of the original attempts by Frege and Dedekind, which were the basis of the later definition by Whitehead and Russell. Nevertheless, his charge of circularity could easily be extended. Indeed, there is a circularity in a definition of the natural number series in terms of the intersection of all the hereditary series (Frege) or chains (Dedekind) to which 0 belongs and to which the successor of x belongs if x belongs to it.[27] Thus the

[25] Indeed, if no satisfactory reinterpretation of the formula defining inductive numbers is possible, the only alternative to rejecting it is an extreme Platonism about the existence of Frege's hereditary properties. This was the position adopted by Ramsey and later by Gödel.

[26] Later on, in one of his last texts, 'La Logique de l'infini', Poincaré summed up the differences of perspective nicely: 'We shall first observe that there are two opposite tendencies among mathematicians in their manner of considering infinity. For some, infinity is derived from the finite; infinity exists because there is an infinity of possible finite things. For others, infinity exists before the finite; the finite is obtained by cutting out a small piece from infinity' (Poincaré 1963: 66).

[27] Frege introduced the notion of hereditary series in the third part of his *Begriffsschrift* (1967), and uses it to justify mathematical induction in his *Foundations of Arithmetic* (1980a: §§ 79–80). Dedekind's definition in terms of chains (*Ketten*) is

totality to be defined is presupposed: in such definitions, the totality of the natural numbers is needed among the totality of all the hereditary series, chains, or inductive classes; the definiendum is already contained in the definiens. Logicists wanted to avoid any reference to the 'and so on' in their definitions, as we can see from a remark by Russell in his *Introduction to Mathematical Philosophy*:

> What are the numbers that can be reached, given the terms '0' and 'successor'? Is there any way by which we can define the whole class of such numbers? We reach 1, as the successor of 0; 2, as the successor of 1; 3, as the successor of 2; and so on. It is this 'and so on' that we wish to replace by something less vague and indefinite. (Russell 1919: 20–1)

This remark is rather important, because the expression 'and so on' that Russell wishes to eliminate is precisely the reference to the process of iteration which characterizes in the eye of constructivists the series of natural numbers. It is precisely in trying to eliminate this 'and so on', and therefore any reference to the potentiality of the processes involved, that Dedekind, Frege, and Russell produced their circular definitions. It is now customary wisdom to consider these second-order definitions as impredicative,[28] and no weakening of the definitions seems really to succeed in avoiding impredicativity.[29]

found in Dedekind (1963b: §§ 37, 44, 59, 60, 80), and an informal presentation is found in Dedekind (1967), where he recognizes the similarities between Frege's notion and his own.

[28] See e.g. Parsons (1983: 132–3).

[29] There have been attempts at giving a weak second-order definition but, according to Daniel Isaacson, such a definition 'does not fare significantly better on the score of avoiding impredicativity than the one based on full second-order logic' (1987: 156). For example, Hao Wang reported in 'Eighty Years of Foundational Studies' a definition suggested by Michael Dummett, in discussion, which reads as follows (Wang 1958: 491):

$$Ny =_{df} (\forall X)\ [(0 \in X \wedge (\forall x)\ (x \neq y \rightarrow Sx \in X) \rightarrow y \in X] \\ \wedge (\exists X)\ (0 \in a \wedge (\forall x)\ (x \neq y \rightarrow Sx \in X)\ )$$

Alexander George has also recently given (George 1987: 515) a revised version of a definition originally given by W. V. Quine :

$$Nx =_{df} (\forall X)\ (x \in X \wedge (\forall y)\ (Sy \in X \rightarrow y \in X) \rightarrow 0 \in X) \\ \wedge (\exists X)\ (x \in X \wedge (\forall y)\ (Sy \in X \rightarrow y \in X)$$

Both definitions are similar in that they are both predicative if the variable X ranges over finite collections only; otherwise they would simply fail to pick up just the natural numbers. Both definitions are not impredicative in the narrow sense of the word, according to which the set defined is not required to lie within the range of the quantifier '∀X' of the definition. But in a less restricted sense of impredicativity, these definitions remain impredicative. Indeed, as was pointed out to me by Daniel Isaacson, in order to understand these definitions one needs to see that '∀X' contains an

Although Russell did not adhere to Poincaré's predicativism, he was deeply influenced by his criticisms: he came to think that the source of the paradoxes was impredicativity and he turned his 'Vicious-Circle Principle', according to which 'Whatever involves *all* of a collection must not be one of the collection', into the cornerstone of the theory of types (*PM* i, p. 37). He also felt compelled to use a notion of 'predicative function' which was considered later, by Ramsey in particular, to be too narrow. Ramsey's stance was that if one adheres to the actual infinite to begin with—as Poincaré had already remarked—then there is no reason to entertain ontological qualms of the Russellian sort: room is already made for a more extreme form of Platonism, according to which all extensions are already pre-existing to our constructions. From this standpoint, the impredicativity of our definitions is the consequence of our human limitations and there is nothing vicious with circular definitions. (The problem of our human limitations, as opposed to the infinite power of God, is lurking in the background; it will be discussed further in section 6.3.) But Poincaré's predicativism did not influence only Russell, albeit indirectly; it was also taken very seriously by Hermann Weyl in his first major work on foundations of mathematics in 1918, *The Continuum*:

> I became firmly convinced (in agreement with Poincaré, whose philosophical position I share in so few other respects) that *the idea of iteration, i.e., of the sequence of the natural numbers, is an ultimate foundation of mathematical thought*—in spite of Dedekind's 'theory of chains' which seeks to give a logical foundation for definition and inference by complete induction without employing our intuition of the natural numbers . . . Moreover, I must find the theory of chains guilty of a *circulus vitiosus*. (Weyl 1987: 48)[30]

Wittgenstein also spoke of circularity with regard to the Frege–Russell definition of natural numbers.[31] We saw in section 2.1 that Wittgenstein's *Formenreihe* are written:

isomorphic copy of the natural number series, i.e. sets of finite ordinals: {0}, {0, 1}, {0, 1, 2}, {0, 1, 2, 3}, . . .

[30] See also Weyl (1968: ii. 149; 1959: 48–9). Weyl's predicativist programme in *The Continuum* was developed further by Paul Lorenzen in *Einführung in der operative Logik* (1955), and it was studied extensively from a proof-theoretical point of view by Georg Kreisel (1960) and Solomon Feferman (1964; 1968).

[31] There is no evidence that Wittgenstein ever read Poincaré, but he probably became acquainted with Poincaré's ideas while studying with Russell before the war.

$\exists x$ (*a*R*x* & *x*R*b*)
$\exists x, y$ (*a*R*x* & *x*R*y* & *y*R*b*)
and so on

(*TLP*, 4.1252, 4.1273), and that the general form of such series is rendered by the variable of 5.2522, [*a*, *x*, *O'x*]. We also saw how Wittgenstein's expression of the general form of integers in 6.03 was closely modelled on this 'variable', and that the parallel with the arithmetical iteration functional (and thus with lambda-calculus) is striking. It is clear that Wittgenstein intended to take as a primitive the notion of recursive or *iterative* process. This approach is in marked contrast with that of logicists such as Frege and, later, Russell, who tried to explain the notion of 'successor of' in terms of the ancestral relation, which is known in *Principia Mathematica* as R*. Using the ancestral relation, Russell was able to provide an explicit definition of the natural numbers series; but it required a set-theoretic apparatus (for one-to-one correlations) which was not available to Wittgenstein. Consequently, he had to reject such explicit definitions, which he did, using an argument parallel to Poincaré's. It is in fact difficult to imagine that Wittgenstein did not have Poincaré's remarks in mind when he criticized the Frege–Russell definition of the ancestral relation:

> If we want to express in logical symbolism the general proposition 'b is successor of a' we need for this an expression for the general term of the formal series: aRb, ($\exists$x) : aRx. xRb, ($\exists$x,y): aRx . xRy . yRb, . . . The general term of the formal series can only be expressed by a variable, for the concept symbolized by 'term of this formal series' is a formal concept. (This Frege and Russell overlooked; the way in which they express general propositions like the above is, therefore, false; it contains a vicious circle.) (*TLP*, 4.1273)

According to Elizabeth Anscombe, the accusation of circularity was a 'peculiarly vicious blow' to Russell: a circularity was found in *90.163, *90.164 and in the proof of *90.31 (Anscombe 1971: 128). The second edition of *Principia Mathematica* included an appendix in which this objection was dealt with (*PM* i, pp. 650–8); but this renewed attempt at predicatively defining the set of natural numbers was also a failure.[32] It is worth noticing that Wittgenstein's diagnosis of the circularity is

[32] For the result that 'the property of being a natural number (in Russell's sense) is not predicatively definable from $\in$ and = taken as primitive', see Myhill (1974: 27). Cf. Leblanc (1975).

different from the usual one: Frege and Russell are said to have 'overlooked' the fact that the concept 'term of this formal series' is a 'formal' concept. In the *Tractatus Logico-Philosophicus* such concepts, which are also called 'logical' or 'pseudo-concepts', can be represented only by a 'variable' and not as a propositional function (*LWL*, pp. 10–11). This is not the case in Frege's and Russell's systems, hence the 'vicious circle' in their definition.

That Wittgenstein, in contrast to the logicists, would not eliminate the 'and so on' can already be seen from the 'variable' expressing the general form of integers in 6.03: $[0, \xi, \xi + 1]$. We also saw in section 2.1 that the 'and so on' remains present in the inductive definition of 6.02. The definition sets no limit to the possibility of applying the operation which generates the numbers to the results of its own application. One ought to recall also that the peculiarity of an operation is that the result of applying it may be used in turn as the basis for another application of it (*TLP*, 5.251), and that this is the only way to progress in a *Formenreihe* (*TLP*, 5.252). Wittgenstein called the repeated application of an operation to its own result its 'successive application' (*TLP*, 5.2521); and '[t]he concept of the successive application of an operation is equivalent to the concept "and so on" ' (*TLP*, 5.2523).

In 1923 Ramsey visited Wittgenstein in the Austrian village of Puchberg am Schneeberg, where the latter was a school teacher. On that occasion, Wittgenstein annotated his copy of the *Tractatus Logico-Philosophicus*. Some of the marginal remarks[33] leave no doubt about Wittgenstein's intentions. In the margins of the page 155, Wittgenstein wrote that 'the fundamental idea of math. is the idea of *calculus* represented here by the idea of *operation*',[34] and that 'The beginning of logic presupposes *calculation* and so number' (Lewy 1967: 421–2). This last remark should help to correct a typical misreading of Wittgenstein's claim in 6.2 (repeated in 6.234) that 'Mathematics is a logical method'. This claim led members of the Vienna Circle among others to believe that Wittgenstein was claiming that mathematical equations are tautologies. Nothing is further

[33] These annotations are reported in Lewy (1967).

[34] In the right-hand margin next to 6.02, Wittgenstein also wrote: 'Number is *the* fundamental idea of calculus and must be introduced as such' (Lewy 1967: 422). But he repudiated this point in *Philosophical Grammar*, where he wrote that '[n]umber is not at all a "fundamental mathematical concept" ' (*PG*, p. 296). One finds the same remark in the *Zettel*: 'Numbers are not fundamental to mathematics' (*Z*, § 706). It is rather the notion of 'operation' which is already in *TLP* the most fundamental one.

from the truth. Thus, one sees not only that the *Tractatus Logico-Philosophicus* contains a series of criticisms of the system of *Principia Mathematica*, including a variant of Poincaré's charge of circularity, but that the very spirit of Wittgenstein's 'reduction' was antithetical to logicism. Differences between Wittgenstein and logicists such as Russell run very deep.[35]

Perhaps it is relevant to say a word or two about other responses to the problem of impredicativity. Ramsey's discussion is particularly important; we will encounter it on many occasions in later sections. In his 1926 paper on 'The Foundations of Mathematics', he argued that Russell gave too much importance to the Vicious Circle Principle. Ramsey could not see, prima facie, that impredicativity was problematic since there is, for example, nothing problematic in referring to a man as the 'tallest in a group', even though that man would then be identified by means of a totality of which he is himself a member (Ramsey 1978: 192). The domain being, in this particular example, finite, impredicativity poses no serious problem. But Ramsey would claim that there is nothing wrong in adopting the same attitude in the infinite case. He could only maintain this, however, by adopting an extreme form of Platonism about existence according to which all properties and their extensions already exist prior to our constructions or descriptions. The fact that some properties can only be described impredicatively is thus seen as caused by 'our inability to write propositions of infinite length', and Ramsey insisted that this inability is 'logically a mere accident' (1978: 192).

This argument played a crucial role for Ramsey, since it ultimately allowed him to use a simple theory of types, thus getting rid of ramification and the controversial Axiom of Reducibility. (For further details, see section 3.2.) This solution met initially with resistance from members of the Vienna Circle, who voiced their objections at the Königsberg meeting in 1930: Rudolf Carnap regarded Ramsey's

[35] Supplementary evidence for this reading of Wittgenstein's critique of the logicist reduction of the principle of mathematical induction can be garnered from Friedrich Waismann's *Introduction to Mathematical Thinking*, whose original version dates from the mid-1930s. In ch. 8, Waismann criticizes the logicist reduction of the principle of mathematical induction along similar lines, claiming that his critique is the vindication of Poincaré's remark that 'the principle of induction is not demonstrable in a logical way' (Waismann 1951: 98). In the Epilogue Waismann mentions that, among other ideas, the views on induction expressed in ch. 8 were taken from an 'unpublished manuscript of Ludwig Wittgenstein which he has been allowed to peruse' (p. 245). (The MS must have been an early one (1929–31), possibly that which was published under the title *Philosophical Remarks*.)

position as 'theological mathematics' (1983: 50), while in his contribution to the discussion, Hans Hahn regarded it 'as impossible' (1980: 35). Carnap and Felix Kaufmann tried to find a solution by reinterpreting the Fregean definition of the natural numbers using Kaufmann's distinction between 'individual' (Carnap used the term 'numerical') and 'specific' generality. According to them, the belief that one must survey all the elements (individuals) of the domain in order to verify a generalized statement is a confusion between these two kinds of generality (Kaufmann 1930: 71, 76; Carnap 1983: 51). 'Specific' generality is not to be understood as established by verifying it for all the individuals; it is somewhat 'analytic': 'We do not establish specific generality by running through individual cases but by logically deriving certain properties from certain others' (Carnap 1983: 51). Gödel, who reviewed Carnap's paper in 1932 (Gödel 1986: 245), pointed out twelve years later in 'Russell's Mathematical Logic' that Carnap's solution, even if it is not without difficulties,[36] has the advantage of getting rid of the circularity (Gödel 1983: 455–6). At any rate, Gödel did not need Carnap's solution, since he opted for an extreme Platonism close to that of Ramsey. Gödel clearly saw that there was a problem with the classical account of universal quantification as an infinite logical product and with the inductive definitions of numbers, but he claimed that the problem arises only for those adopting the 'constructivist standpoint' according to which mathematical entities are constructed by ourselves. For those not holding such a viewpoint, there are no problems:

> even if 'all' means an infinite conjunction, it seems that the vicious circle principle . . . applies only if the entities involved are constructed by ourselves . . . If . . . it is a question of objects that exist independently of our constructions,

[36] See Gödel (1983: 457–8). The problem with Carnap's solution is that in order to establish that all truths of mathematics are analytic, he had to face Gödel's incompleteness theorems, and he could not obtain his result from deduction from any usual axiom system. This is why he used a version of the ω-rule, which is rule DC2 in *The Logical Syntax of Language* (Carnap 1937: 38). Adding this rule to an axiom system strong enough to represent Peano Arithmetic renders this system complete with respect to all truths expressible in the language of arithmetic. (See Theorem 14.3 in *The Logical Syntax of Language* (Carnap 1937: 40) for Carnap's version of this result.) The ω-rule is therefore a useful tool to extend provability beyond Peano Arithmetic, but it requires for its justification acceptance of modes of argument which lie beyond those usually needed for Peano Arithmetic, and any weakened version of the ω-rule which could be justified by these does not extend provability beyond Peano Arithmetic. The aim of Carnap's definition of analyticity is thus defeated. On these questions, see Daniel Isaacson's excellent discussion in Isaacson (1991).

there is nothing in the least absurd in the existence of totalities containing members, which can be described (i.e., uniquely characterized) only by reference to this totality. (Gödel 1983: 456)

As a solution, Gödel adopted an extreme form of Platonism, assuming that mathematical objects exist independently of us and that they have a reality as legitimate as that of physical bodies (1983: 456). In this form of Platonism, the circularity of the definitions is not denied: it is only claimed that it causes no problem. Circularity is thus seen as problematic only for those who adopt a 'constructivistic' or 'nominalist' stance—these are Gödel's own words—and even *Principia Mathematica* appears, from this angle, to be 'constructivistic' (Gödel 1983: 456).[37] From Gödel's point of view Ramsey was ultimately a prisoner of the theory of types: although he showed how to reformulate it in order to avoid the Axiom of Reducibility, he had, in Gödel's words, to make 'the fiction that one can form propositions of infinite and even non-denumerable length, i.e., [operate] with truth-functions of infinitely many arguments, regardless of whether or not one can construct them' (Gödel 1983: 460–1). For Gödel, such infinite truth-functions were needlessly complicated objects to deal with (as opposed to classes) and made little sense, since they could be understood only by an infinite being (p. 461).[38] All this was for Gödel a 'verification', once more, of his view that 'logic and mathematics (just as physics) are built up on axioms with a real content which cannot be "explained away" ' (p. 461). We shall see in sections 3.2 and 4.2 that Wittgenstein was to react negatively to Ramsey's introduction of infinite truth-functions, thus moving towards a more finitist standpoint. (Even Ramsey himself ultimately abandoned this point of view.) So we can see that in the end the problem of predicativity led to a radicalization of the positions ranging from the predicativism of Poincaré and Weyl, with which Wittgenstein could be (somewhat inappropriately) grouped, to the extreme Platonism of Gödel, while the positions of Russell (and with it Ramsey's more radical position, with the commitment to infinite truth-functions) and Carnap being considered inadequate and abandoned. Wittgenstein's viewpoint and Gödel's view that mathematics is built up 'on axioms with a real content' are, so to speak, at opposite ends of the spectrum.

[37] Of course, this 'constructivistic' approach is, so to speak, watered-down in *Principia Mathematica*, where the Axiom of Reducibility and the Axiom of Infinity make up for arbitrary functions, etc. See Gödel (1983: 461).

[38] For attempts at an infinitary logic, see Scott and Tarski (1958); Karp (1964).

3

Arbitrary Functions

3.1. IDENTITY

In contemporary mathematical logic, the debate over the notion of arbitrary function presented in Chapter 1 is reflected in the problem of the interpretation of second-order quantifiers. The distinction between first- and second-order logic hangs on the range of the quantifiers: in first-order logic, quantifiers range uniquely on elements of the 'structure' under study, while in second-order logic, quantifiers can range over the latter's subsets, sets of subsets, and so forth. The range of second-order quantifiers admits of various interpretations from more encompassing or 'standard' ones to more restrictive or 'non-standard' ones:[1] those adopting the standard interpretation would claim that the range of a second-order quantifier involving, say, a one-place class variable X whose values are classes of individuals of a domain do(M) is the entire power set *P*(do(M)), so that some values of X are arbitrary extensionally possible classes, while those adopting the non-standard interpretation would consider only some such classes as constituting the range of the quantifier. The same reasoning applies if X is a predicate variable or a function variable (especially when one looks at functions as sets of ordered pairs).

When Leon Henkin made this distinction explicitly in his paper on the 'Completeness in the Theory of Types', he considered only one such non-standard interpretation, with his general models, where the higher-order variables are subjected to closure conditions with respect to Boolean and projective operations (Henkin 1950). But there are different kinds of non-standard interpretation, such as Gödel's so-called *Dialectica* interpretation, where he restricted the value of the function variables to recursive functions of the appropriate type (Gödel 1980). The *Dialectica* interpretation is in a weak sense constructive, and although one ought not to confuse non-standard inter-

[1] As a matter of convenience, I shall adopt here the terminology standard/non-standard which has been used e.g. by Jaakko Hintikka (1995a).

pretations with constructivism (there is certainly nothing constructive with Henkin's general models), various constructivist programmes can be seen as the adoption of a non-standard interpretation. Careful use of the standard/non-standard distinction may thus help bring to the fore neglected aspects of foundational disputes such as that over Dirichlet's notion of an arbitrary function:[2] the distinction between those objecting to the general notion of arbitrary function and those adopting it and the distinction between the standard and the non-standard interpretations overlap considerably, since the standard interpretation of higher-order quantifiers is, for our purposes, practically equivalent to the notion of an arbitrary function. We can thus see, for example, that Kroneckerian-minded French mathematicians (Baire, Borel, and Lebesgue) were opting for a non-standard interpretation, and that the predicativism of Poincaré is an even more radical non-standard interpretation.

This dispute provides the appropriate background against which one may study Wittgenstein's remarks on a number of topics, such as identity, Ramsey's 'functions in extension', the Axiom of Infinity, the Axiom of Choice, and numerical equivalence. We have already seen that Wittgenstein's standpoint in the *Tractatus Logico-Philosophicus* coheres to some degree with Poincaré's radical form of definabilism, since proposition 4.1273 presents, as we saw in section 2.3, a variant of Poincaré's argument of circularity directed at the core of *Principia Mathematica*. But this is not the only aspect of the *Tractatus Logico-Philosophicus* which points to a radical form of non-standard interpretation: his rejection of the notion of identity as a relation is as clear an indication as anything else that he took such a stance.[3]

[2] The distinction between the standard and non-standard interpretations of second-order quantifiers is thus a useful tool for investigations in the history of the foundations of mathematics. But there are some obvious dangers here, since early pioneers such as Frege did not distinguish sufficiently between first- and second-order logic for them to be properly aware of the distinction. Therefore, any reading of the standard/non-standard distinction in the writings of such authors will involve some amount of distortion. The attribution of the non-standard interpretation to Frege by Hintikka and Sandu, by virtue of a primacy of concepts over their extensions (Hintikka and Sandu 1992: 298 ff.), has provoked a heated debate. For replies to Hintikka and Sandu, see Burgess (1993); Demopoulos and Bell (1993); Heck and Stanley (1993).

[3] In what follows I shall limit myself to a discussion of Wittgenstein on identity from the angle of the foundational issue about arbitrary functions. There is no claim that my discussion exhausts the issues raised by Wittgenstein's remarks. The reader is referred to Roger White's excellent study, 'Wittgenstein on Identity' (1979).

Since Frege's *Begriffsschrift*, the sign for equality, =, has been considered by logicians as a logical constant. It is also usually assumed that mathematical equality is only a special case of the general concept of logical identity,[4] but Wittgenstein sharply distinguished between these two notions,[5] and he argued for the elimination of logical identity (*TLP*, 5.53 ff.), while still being able to hold the view that mathematics consists of equations. Within the context of the *Tractatus Logico-Philosophicus*, the obvious motivation for this move is that Wittgenstein had no choice but to explain away the identity sign in order to carry on with his theory of logical truth according to which every proposition is a truth-function of elementary propositions (*TLP*, 5). Wittgenstein used the N operator to show that theorems of the predicate calculus are tautologies, and he had to show that logical truths such as

$\exists x\ (x = x)$, $\forall x\ (x = x)$, etc.

were either reducible to tautologies or ruled out as nonsense.

Wittgenstein wrote that 'one cannot e.g. say "There are objects" as one says "There are books". The same goes for "There are 100 objects" or "There are $\aleph_0$ objects"' (*TLP*, 4.1272). This is because the expression 'object' (along with other expressions such as 'number') stands for a logical form, not a concept. It designates a 'pseudo-concept' (*Scheinbegriff*) that can be properly expressed only by a variable, and the fact that there are 'objects' or 'numbers' *shows* itself by the employment of suitable variables in the formal language. Thus propositions such as 'There are objects', where the word 'object' is used as a proper concept word (*eigentliches Begriffswort*), could not be written down in a form using the identity sign, as above, and therefore could not be conceived as truth-functions of elementary propositions. These were meant to picture the way objects are disposed to one another, and Wittgenstein could not see identity as a relation between objects: as a consequence of the ineffability of the name–object relation, the existence of an individual can only be shown through the use of its name: 'Identity of the object I express by identity of the sign and not by means of a sign of identity. Difference of the objects by difference of the signs' (*TLP*, 5.53). His

[4] See e.g. Tarski (1965: 61).

[5] See e.g. *WVC*, p. 146. This was not and still is not the received view. It seems, if this can be any consolation, that W. E. Johnson agreed with Wittgenstein on that point (Johnson 1922–4: ii. 143).

argument against the notion of identity as a relation between objects is very simple and powerful: on the one hand it is indeed nonsense to say of two different individuals that they are the same; to say, on the other hand, of one individual that it is identical to itself is to say nothing (*TLP*, 5.5303).

In *Principia Mathematica*, Russell defined identity as follows: $x = y$ if and only if x and y satisfy exactly the same predicative functions. This is definition *13.01:

(1) $(x = y) \leftrightarrow \forall F\ (F(x) \rightarrow F(y))$

which retains much of the flavour of Leibniz's law of the identity of indiscernibles. Against this definition, Wittgenstein pointed out that although it may never be true, it nevertheless makes sense to suppose that two different objects x and y share all their properties. This is ruled out by Russell's definition, however, so the latter must be wrong: 'Russell's definition of "=" won't do; because according to it one cannot say that two objects have all their properties in common. (Even if this proposition is never true, it is nevertheless *significant*.)' (*TLP*, 5. 5303).[6] Ramsey used the same argument against Russell in 'The Foundations of Mathematics':

> we ought not to define identity in this way as agreement in respect of all predicative functions, because two things can clearly agree as regards all atomic functions and therefore as regards all predicative functions, and yet they are two things and not, as the proposed definition of identity would involve, one thing. (Ramsey 1978: 201)

As we shall see, the difference was that Ramsey tried to provide an extension of the notion of propositional function in order to provide for classes that cannot be defined by predicative functions, while these were simply excluded by Wittgenstein.

In the *Tractatus Logico-Philosophicus*, the elimination of the identity sign is matter of simple transformations.[7] These are best seen by using an example (*TLP*, 5.532): the expression 'There are two objects which have the property F', which is usually rendered by the formula

[6] Wittgenstein gave a rather similar argument in *TLP*, 5.5352: 'Similarly it was proposed to express "there are no things" by "$\neg\ (\exists x).\ x = x$". But even if this were a proposition—would it not be true if indeed "There were things", but these were not identical with themselves?'. Indeed, $\neg\ \exists x\ (x = x)$ could be taken as meaning that nothing is self-identical, and this is consistent with there being some non-self-identical things.

[7] For details, see Hintikka (1956).

$$\exists x\, \exists y\ (F(x) \wedge F(y) \wedge (x \neq y))$$

should be rewritten (because the variable names x and y are here the 'proper sign[s] for the pseudo-concept *object*' (*TLP*, 4.1272)) as

$$\exists x\, \exists y\ (F(x) \wedge F(y))$$

or, following 5.532, as

$$\exists x, y\ F(x, y)$$

which is to be true if and only if $F(x,y)$ is true for some substitution for x and for some different substitution for y. This is the crucial point. Recall that in the *Tractatus Logico-Philosophicus* quantifiers are interpretable as logical sums and products (such an interpretation being, as we saw in section 2.2, not exactly faithful to the spirit of the *Tractatus*). One can now see that Wittgenstein replaced identity by a convention for the reading of the quantifiers or, rather, a new interpretation of variables which is in accordance with his idea that the identity of object should be expressed by the identity of sign. According to Alice Ambrose's lecture notes, Wittgenstein gave this explanation: 'we can make it a rule not to write signs of equality, but instead write one variable if one wants to talk of exactly one thing, two if one talks of two things' (*AWL*, p. 146). According to Wittgenstein's reading, coincidences of the values of different variables are excluded. This reading was named by Jaakko Hintikka the 'exclusive interpretation of variables' (Hintikka 1956: 226). This peculiar interpretation explains Wittgenstein's unorthodox way of writing more than one variable next to the quantifier signs and the function sign F. Wittgenstein's interpretation is clumsier, since, for example,

$$\exists x\ (F(x) \wedge \exists y\ F(y) \wedge F(x))$$

is reducible to

$$\exists x\ F(x)$$

in the ordinary, 'inclusive' interpretation of the variables but not in Wittgenstein's, where the coincidences between the bound variables x and y are excluded (Hintikka 1956: 229).

Wittgenstein's elimination of identity leads directly to an important criticism of the theory of cardinal numbers in *Principia Mathematica*. In order to show this, we need go no further than the introduction of the Unit Class in *52 and of cardinal couples in *54. The cardinal number 2 is defined in *54.02 as the class of all couples of the form

$$\iota\text{'}x \cup \iota\text{'}y$$

(with $x \neq y$), the latter being defined as 'the class whose only members are x and y', since if there is a third term z in $\iota\text{'}x \cup \iota\text{'}y$, then $(z = y) \vee (z = y)$ (*51.232). So, couples are of the form:

$$((x = a) \wedge (y = b)) \vee ((x = c) \wedge (y = d)) \vee \ldots$$

(Of course, here the sign = does not express mathematical equality but logical identity.) Wittgenstein was extremely annoyed with Russell's criterion for 'sameness of number' or numerical equivalence—he never ceased to criticize this notion—because as he understood it 'no correlation seems to be made' (*AWL*, p. 149). Once again, this annoyance comes from Wittgenstein's deep-rooted conviction of the non-standard interpretation: if one cannot provide any material correlation—as in the case of correlating cups and spoons to be presented below in section 3.3—one must provide a rule for the correlation; there should be no roundabout way to avoid providing it. But this is precisely what he took Russell to be doing: 'Russell had a way of getting round this difficulty. No correlation need actually be made, since two things are always correlated with two others by identity' (*AWL*, p. 149). Wittgenstein's reasoning, according to Ambrose's lecture notes, is that we have here two functions satisfied only by two individuals each, respectively a,b and c,d:

$$((x = a) \vee (x = b)) \text{ and } ((y = c) \vee (y = d))$$

If one substitutes a for x and c for y one obtains:

$$((a = a) \vee (a = b)) \text{ and } ((c = c) \vee (c = d))$$

From this one can construct a function satisfied only by ac and bd,

$$((x = a) \wedge (y = c)) \vee ((y = b) \wedge (x = d))$$

or

$$((x = a) \vee (y = d)) \wedge ((x = b) \vee (y = c))$$

These functions correlate a and c and b and d by mere identity. Wittgenstein's conclusion is that 'if "=" makes no sense, then it is no correlation. Why does this function seem to correlate them? Because of the identity sign' (*AWL*, p. 150). As we shall see in the next section, Wittgenstein raised a similar objection to Ramsey's functions in extension, in his *Philosophical Grammar*.[8] This is precisely the kind of consequence of the elimination of identity which led Wittgenstein to talk of the theory of classes as 'superfluous' (*TLP*, 6.031). As we shall see in the next section, this was a subject of grave concern for Ramsey. It seems that the theory of cardinal numbers simply cannot get off the ground. The rejection of identity in the *Tractatus Logico-Philosophicus* leads to a rejection of the notion of numerical equivalence, which is at the core of the logical constructions of Frege and Russell. This much is implicit in the *Tractatus Logico-Philosophicus*, but later Wittgenstein felt compelled to support his standpoint by providing an argument directed against the notion of numerical equivalence. It will be examined in section 3.3 below.

There are other important consequences to the rejection of identity as a relation, which were also immediately noted by Wittgenstein: 'And we see that apparent propositions like: "$a = a$", "$a = b.b = c.\supset a = c$", "$(x).x = x$", "$(\exists x).x = a$", etc. cannot be written in a correct logical notation at all' (*TLP*, 5.534). One will recognize in this list some of the theorems of the theory of identity in *Principia Mathematica*. For example,

$$x = x$$

(reflexivity of identity or 'everything is equal to itself') is theorem *13.15, and

$$((x = y) \wedge (y = z)) \rightarrow (x = z)$$

(transitivity of identity) is theorem *13.17. The list is of course incomplete. Wittgenstein could have added symmetry, for example, i.e. theorem *13.16:

$$(x = y) \rightarrow (y = x)$$

Leibniz's law of the identity of indiscernibles also becomes a proposition with a sense; it is not a tautology. It is thereby thrown out of predicate logic. That such existence propositions cannot attain the status of theorems in the *Tractatus Logico-Philosophicus* is not the only important consequence: Russell's Axiom of Infinity is also jeopardized, and while Wittgenstein was happy with this consequence, it

[8] There is also an argument similar to the one just presented in *PG*, pp. 355–6.

was the subject of concern to Ramsey, as we shall see in the next section.

3.2. RAMSEY'S FUNCTIONS IN EXTENSION AND THE AXIOM OF INFINITY

When Ramsey visited Wittgenstein in Austria in 1923, he had already written a review of the *Tractatus Logico-Philosophicus* for *Mind* (Ramsey 1923) which still stands as one of the better commentaries on this difficult book. But he nevertheless wanted Wittgenstein to clarify for him some of its most difficult passages. However, having it thoroughly explained to him was not the only purpose of Ramsey's visit. He wrote to his mother that 'when the book is done I shall try to pump [Wittgenstein] for ideas for its further development which I shall attempt' (*LO*, p. 74). Russell was also preparing at that time the second edition of *Principia Mathematica*, and Ramsey reported in the same letter that Wittgenstein was 'a little annoyed that Russell is doing a new edit[ion] of Principia because he thought he had shown R[ussell] that it was so wrong that a new edition would be futile. It must be done altogether afresh' (*LO*, p. 74). A fresh start was clearly what Ramsey thought he was making during the following year, while writing his paper on 'The Foundations of Mathematics' (Ramsey 1978: 152–212). It was his intention to render *Principia Mathematica* 'free from the serious objections which have caused its rejection by the majority of German authorities' by 'using the work of Mr. Ludwig Wittgenstein' (p. 152). At any rate, it seems that Ramsey did help Russell with the second edition of the *Principia Mathematica*, probably by trying to explain Wittgenstein's criticisms,[9] but the ultimate results certainly disappointed him. Already in February 1924, Ramsey had seen the manuscript of the second edition and wrote to Wittgenstein: 'You *are quite right that it is of no importance*; all it really *amounts to is a clever proof of mathematical induction without using the axiom of reducibility*. There are no fundamental changes, identity just as it used to be' (*LO*, p. 84). (As we saw in section 2.3, Russell's new proof was not so 'clever'.)

[9] It is stated in the introduction to the 2nd edn. that 'the authors are under great obligations to Mr F. P. Ramsey of King's College, Cambridge, who has read the whole in MS. and contributed valuable criticisms and suggestions' (*PM*, p. xiii n.).

In 'The Foundations of Mathematics', Ramsey was to point out three 'great defects' in *Principia Mathematica*. The first has to do with Ramsey's thesis that mathematics is 'essentially extensional' (Ramsey 1978: 166). By 'calling mathematics extensional' Ramsey meant that 'it deals not with predicates but with classes, not with relations in the ordinary sense but with possible correlations, or "relations in extension" as Mr Russell calls them' (p. 165). He believed that infinite indefinable classes (arbitrary subsets of an infinite set) are essential to modern mathematics[10] and he was aware of the limitations of *Principia Mathematica* in that respect. Indeed, one of the best and earliest anticipations of Henkin's standard/non-standard distinction is to be found in 'The Foundations of Mathematics':

> The theory of *Principia Mathematica* is that every class or aggregate (I use the words as synonyms) is defined by a propositional function—that is consists of the values of x for which 'ϕx' is true, where 'ϕx' is a symbol which expresses a proposition if any symbol of appropriate type be substituted for 'x'. This amounts to saying that every class has a defining property. Let us take the class consisting of a and b; why, it may be asked, must there be a function $\phi \hat{x}$ such that 'ϕa' and 'ϕb' are true, but all other 'ϕx's false? This is answered by giving as such a function '$x = a \,.\, \vee \,.\, x = b$'. Let us for the present neglect the difficulties connected with identity, and accept this answer; it shows us that any finite class is defined by a propositional function constructed by means of identity; but as regards infinite classes it leaves us exactly where we were before, that is, without any reason to suppose that they are all defined by propositional functions, for it is impossible to write down an infinite series of identities. To this it will be answered that a class can only be given to us either by enumeration of its members, in which case it must be finite, or by giving a propositional function which defines it. So that we cannot be in any way concerned with infinite classes or aggregates, if such there be, which are not defined by propositional functions. But this argument contains a common mistake, for it supposes that, because we cannot consider a thing individually, we can have no concern with it at all. Thus, although an infinite indefinable class cannot be mentioned by itself, it is nevertheless involved in any statement beginning 'All classes' or 'There is a class such that', and if indefinable classes are excluded the meaning of all such statements will be fundamentally altered. (Ramsey 1978: 173)

This passage—especially the last sentence—is a clear indication that Ramsey thought that if one admits only infinite classes definable

[10] It is perhaps worth mentioning that the only other contemporary Cambridge philosopher agreeing with Ramsey on the need for, in his own words, 'factitious correlations', was W. E. Johnson, in his *Logic* (1922: ii. p. 159).

by propositional functions, as in *Principia Mathematica*, the interpretation of the higher-order quantifiers will be altered. In short, Ramsey thought that Whitehead and Russell had mistakenly adopted in *Principia Mathematica* what amounts to a non-standard interpretation.[11] The defect was thus that, by considering only the infinite classes which are defined by a propositional function, Whitehead and Russell provided too narrow a frame to account for modern mathematics:

> The mistake is made not by having a primitive proposition asserting that all classes are definable, but by giving a definition of class which applies only to definable classes, so that all mathematical propositions about some or all classes are misinterpreted. (Ramsey 1978: 174–5)

I shall not discuss here the merits of Ramsey's claim; it would lead us too far afield.[12] The most unlikely consequence of Whitehead and Russell's narrow definition of predicative function is the introduction of the Axiom of Reducibility, which says (in a simplified form) that for every propositional function ϕ there is some predicative function ψ which is satisfied by all the same arguments and no other:

$$\forall\phi\ \exists!\psi\ (\phi(x) \leftrightarrow \psi(x))$$

(It is worth noticing here the link between this axiom and the definition of identity in definition *13.01, which was discussed in the previous section: the latter is based on the former.) This axiom, needed in order to re-establish the standard interpretation, is counter-intuitive and was considered by all as a blemish. In the preface to the second edition of *Principia Mathematica*, Russell could only give a 'purely pragmatic justification'.[13] He pointed out a number of crucial theorems obtained only with its addition to the system: 'it leads to the desired results, and to no others' (*PM* i, p. xiv).

In his paper, Ramsey tried to establish a strong form of standard interpretation first by redefining the notion of predicative function in

[11] One may think that the Axiom of Reducibility was precisely devised to bring about the standard interpretation. It is not clear, however, if it succeeds in doing so. See Hintikka (1995a).

[12] Although his systems of predicative analysis have little to do with *Principia Mathematica*, I ought to mention here Solomon Feferman's study of predicativity, which he began in Feferman (1964; 1968). His recent assessment (1987; 1988; 1993) shows that the belief that a predicativist frame is too narrow is in fact largely unfounded.

[13] See also *PM*, pp. 59–60, where Russell speaks of 'largely inductive' reasons for adopting the Axiom of Reducibility.

such a way that it captures a range of functions large enough—in fact all functions of individuals in *Principia Mathematica* become 'predicative' in Ramsey's sense—to avoid the need for the Axiom of Reducibility and to get rid of qualms about impredicativity (Ramsey 1978: 189–93), and, secondly, by introducing 'impredicative' functions—in Ramsey's sense—with the help of his device of 'functions in extension' (pp. 200 f.). It should be clear that Ramsey was by then moving away from *Principia Mathematica* in a direction opposite to Wittgenstein's. In a nutshell, while the latter tried to eliminate the theory of classes by propounding a reduction to an intensionalist theory of operations, Ramsey wanted to renovate *Principia Mathematica* by extensionalizing it further. A clash would be inevitable; it came over the notion of identity and Ramsey's 'functions in extension'.

Before moving on to the other two major defects of *Principia Mathematica* according to Ramsey, I would like to comment briefly on his definition of a predicative function. According to Ramsey there is an essential difference between functions of individuals and functions of functions which lies in their ranges: while the range of values of a function of individuals is fixed by the range of individuals, which is according to Ramsey 'an objective totality which there is no getting away from' (Ramsey 1978: 187), the range of arguments of a function of functions is the range of symbols which become propositions when the variables they contain are replaced by names for individuals, and 'this range of symbols, actual or possible, is not objectively fixed, but depends on our methods of constructing them' (pp. 187–8). To Russell's method, which he labels 'subjective' and dismisses on the account that it 'leads to the impasse of the Axiom of Reducibility', Ramsey opposed his own 'entirely original objective method':

> My method . . . is to disregard how we could construct (propositions), and to determine them by a description of their senses or imports; and in so doing we may be able to include in the set propositions which we have no way of constructing, just as we include in the range of values of ϕx propositions which we cannot express from lack of names for the individuals concerned. (pp. 187–8)

This method led Ramsey to the definition of a predicative function of individuals as any *finite or infinite* truth-function of either (atomic) functions of individuals or propositions (p. 190). This definition has the advantage mentioned above of capturing enough functions to avoid having recourse to the Axiom of Reducibility. Mention should be made of the fact that it depends on one crucial difference *vis-à-vis Principia Mathematica*, which is precisely the notion of a truth-function with an infinite number of arguments which Ramsey took from the *Tractatus Logico-Philosophicus*. As Ramsey himself hastened to point out, his definition

> is essentially dependent on the notion of a truth-function of an infinite number of arguments; if there could only be a finite number of arguments our predicative functions would be simply the elementary functions of *Principia*. Admitting an infinite number involves that we do not define the range of functions as those which could be constructed in a certain way, but determine them by a description of their meanings. They are to be truth-functions—not explicitly in their appearance, but in their significance—of atomic functions and propositions. In this way we shall include many functions which we have no way of constructing, and many which we construct in quite different ways. (p. 190)

Thus the account of general propositions as logical sums and products of possibly infinite length, taken by Ramsey from the *Tractatus Logico-Philosophicus*, plays an essential role in his redefinition of the notion of predicative function, which is in turn the cornerstone of his programme of renovations for *Principia Mathematica*. This crucial use of infinite truth-functions enabled Ramsey to devise an extreme form of Platonism which amounted to a total abandonment of Russell's 'subjective' standpoint and which allowed him to shed any qualms about impredicativity—this much has already been discussed in section 2.3. We shall come back to Ramsey's argument (concerning our epistemic limitations as opposed to God's infinite powers) in defence of impredicativity in section 6.3, and we shall also see in section 4.1 that Ramsey abandoned this view in favour of a finitist outlook in 1929, the year before his death.

The second defect of *Principia Mathematica*, according to Ramsey, was what he deemed to be a failure to overcome the difficulties raised by the contradictions of the theory of classes. Since these difficulties are not directly linked to those raised by the 'extensionality' of mathematics, I shall not discuss them further.[14] The third major defect was the treatment of identity. This was already hinted at in the letter to Wittgenstein dated February 1924, which was quoted above. Ramsey's objection here was similar to the first:

[14] For details of Ramsey's treatment of the paradoxes, see Sahlin (1990: 160–75).

The real objection to this definition of identity is the same as that urged above against defining classes: that it is a misinterpretation in that it does not define the meaning with which the symbol for identity is actually used. (Ramsey 1978: 181–2)

So it seems that Ramsey agreed with Wittgenstein's conclusion about identity, which was that '[t]he identity sign is therefore not an essential constituent of logical notation' (*TLP*, 5.533). For Ramsey, the fact that identity was eliminable was evidence that Russell's treatment was defective. His endorsement of Wittgenstein's criticisms was without (serious) reservations:

These arguments are reinforced by Wittgenstein's discovery that the sign of identity is not a necessary constituent of logical notation, but can be replaced by the convention that different signs must have different meanings. . . . the convention is slightly ambiguous, but it can be made definite, and is then workable, although generally inconvenient. But even if of no other value, it provides an effective proof that identity can be replaced by a symbolic convention, and is therefore no genuine propositional function, but merely a logical device. (p. 183)

Ramsey did not stop there, however, but introduced a mirror-definition with his notion of 'functions in extension', which was intended to play the role that identity should have played, had it not been misinterpreted in *Principia Mathematica*. Indeed, although Ramsey saw that Wittgenstein's remarks would lead to an improvement of the conceptual notation of *Principia Mathematica*, he also realized that if the latter was shorn of identity, one would be left with insufficient means to define all arbitrary functions:[15]

The difficulty about identity we can get rid of, at the cost of great inconvenience, by adopting Wittgenstein's convention, which unables us to eliminate '=' from any proposition in which it occurs. But this puts us in an hopeless position as regards classes, because, having eliminated '=' altogether, we can no longer use $x = y$ as a propositional function in defining classes. So that only classes with which we are now able to deal are those defined by predicative functions. (Ramsey 1978: 200)

Wittgenstein would have been quite happy, since he did not see any need for the theory of classes. But for Ramsey this was a 'hopeless

[15] See Ramsey (1978: 201). Ramsey had already noticed in his review of *TLP* that the rejection of identity 'may have serious consequences in the theory of aggregates and cardinal number' (Ramsey 1923: 475).

position' precisely because 'mathematics is extensional'. In sharp contrast to Wittgenstein, who wanted none, Ramsey wanted more classes than Russell, so to speak: he wanted to introduce non-predicative propositional functions or, in other words, infinite indefinable classes. Hence the need for the logical device of the 'functions in extension'. We can see behind Ramsey's extensionalism his recognition that one must provide for the largest class of functions, i.e. for arbitrary functions. He wrote that only with the help of his notion of 'function in extension' and of his renovated notion of predicative function would one be able to preserve mathematics from the 'Bolshevic menace of Brouwer and Weyl' (p. 207). Therefore, although Ramsey apparently agreed with Wittgenstein's criticisms of Russell's treatment of identity, his 'functions in extension' were not just a nice way of rewriting propositions involving the sign of identity: with their help, Ramsey thought that he would vindicate the standard interpretation. In this he was going in the opposite direction from Wittgenstein, who could not but react negatively to Ramsey's paper.

Moreover, Wittgenstein's elimination of identity had, as we saw, the effect of ruling out certain expressions as nonsensical, in particular the Axiom of Infinity. This axiom is purely and simply vital to the logicist programme, so Ramsey had to find a way to circumvent Wittgenstein's objections, i.e. to give it some legitimacy. Again, Wittgenstein could not but react negatively. The Axiom of Infinity, which incidentally only appears for the first time in the second volume at *120.03, states that 'if α is any inductive cardinal (finite number), there is at least one class (of the type in question) which has α terms' (*PM* ii, p. 203). It can be read for our purposes as stating that there is an infinity of distinguishable entities. In the *Notebooks*, Wittgenstein had already written that '[a]ll the problems that go with the Axiom of Infinity have already to be solved in the proposition "$(\exists x).\ x = x$" ' (*NB*, p. 10). So when Wittgenstein got rid of $\exists x\ (x = x)$ by eliminating identity in the *Tractatus Logico-Philosophicus*, he wrote:

> So all problems disappear which are connected with such pseudo-propositions.
>
> This is the place to solve all problems which arise through Russell's 'Axiom of Infinity'.
>
> What the Axiom of Infinity is meant to say would be expressed in language by the fact that there is an infinite number of names with different meanings. (*TLP*, 5.535)

Wittgenstein's position is summed up in a letter to Russell:

> A proposition like '$(\exists x). x = x$' is for example really a proposition of physics. The proposition '$(x) : x = x . \supset . (\exists y). y = y$' is a proposition of logic: it is for physics to say *whether any thing exists*. The same holds of the infinity axiom, whether there are $\aleph_0$ things is for experience to settle. (*NB*, p. 128)

Russell himself had some similar qualms about the status of this axiom, perhaps as a result of previous discussions with Wittgenstein. He had already written in *Principia Mathematica* that '[i]t seems plain that there is nothing in logic to necessitate its truth or falsehood, and that it can only be legitimately believed or disbelieved on empirical grounds' (*PM* ii, p. 183).

The Axiom of Infinity remained vital, however, for Ramsey's renovated logicist programme, since nothing short of postulating transfinite types or lifting the prohibition on classes having members of mixed types would avoid the need for the Axiom of Infinity, and these alternatives were not available to Ramsey.[16] Thus he had no choice but to circumvent Wittgenstein's objections while proposing his modifications to *Principia Mathematica*. He extended the Russellian notion of propositional function to produce the following definition: for $x \neq y$, $x = y$ may be taken to be

$$(\exists\phi\ \phi(x) \wedge \neg\ \phi(x)\ ) \wedge (\exists\phi\ \phi(y) \wedge \neg\ \phi(y))$$

(a contradiction) and for $x = y$, $x = y$ may be taken to be

$$(\forall\phi\ \phi(x) \vee \neg\ \phi(x)\ ) \wedge (\forall\phi\ \phi(y) \vee \neg\ \phi(y)\ ),$$

(a tautology) where $x = y$ is not predicative (Ramsey 1978: 202–3). So, according to Ramsey, one had no choice but to introduce non-

[16] Indeed, the first transfinite type will contain the union of all domains of finite type, i.e. it will be infinite. So the natural numbers can be defined at the transfinite level. One needs, however, special axioms for the existence of objects of the transfinite type. To my knowledge, the idea of transfinite type theory was first discussed by Poincaré in 'La Logique de l'infini' (1963: 52), following a suggestion of König, and by Russell, who dismissed the idea because he thought that the number of arguments and apparent variables of a function can only be finite (Russell 1910: 286). The idea was later put forward by Hilbert in 'On the Infinite' (1967b: 387), then referred to by Gödel in his famous 1931 paper on incompleteness (Gödel 1967: 610, n. 48a) and by Tarski in the postscript to 'The Concept of Truth in Formalized Languages' (Tarski 1983: 268 ff.). The first systematic work was done in particular by Maurice L'Abbé (1953) and Peter Andrews (1965). As for the prohibition on classes having members of mixed types, it is lifted in Zermelo–Fraenkel set theory, where the intuitive model includes a hierarchy not unlike that of type theory.

predicative propositional functions by extensionalizing the notion of propositional function:

How is this to be done? The only practicable way is to do it as radically and drastically as possible; to drop altogether the notion that ϕa says about *a* what ϕb says about *b*; to treat propositional functions like mathematical functions, that is, extensionalize them completely. Indeed, it is clear that mathematical functions being derived from propositional, we shall get an adequately extensional account of the former only by taking a completely extensional view of the latter. (p. 203)[17]

In order therefore to replace what had been recognized as a mere 'logical device'—Russell's notion of identity—Ramsey had to introduce another such device, which could not be open to the same criticisms. The device in question was his 'functions in extension', which he introduced as follows:

Such a function of one individual results from any one-many relation in extension between propositions and individuals; that is to say, a correlation, practicable or impracticable, which to every individual associates a unique proposition, the individual being the argument of the function, the proposition its value.

Thus ϕ (Socrates) may be Queen Anne is dead,

ϕ (Plato) may be Einstein is a great man;

$\phi \hat{x}$ being simply an arbitrary association of propositions ϕx to individuals x. (p. 203)

Ramsey wrote $\phi_e \hat{x}$ for a function in extension, and considered the totality of such functions as the domain of the variable ϕ_e. So he replaced Russell's definition (1) in *13.01 by the expression

$$(2) \quad \forall \phi_e \, (\phi_e x = \phi_e y)$$

which asserts that with any function ϕ_e, the proposition correlated with x is equivalent with the proposition correlated with y. This expression is therefore equivalent to the logical product

$$(\phi_e^1 x \equiv \phi_e^1 y) \wedge (\phi_e^2 x \equiv \phi_e^2 y) \wedge (\phi_e^3 x \equiv \phi_e^3 y) \wedge \ldots$$

and if $x = y$, that is if there is a real identity between x and y, the propositions correlated will always come out the same, that is

[17] Compare Ramsey's words: 'to drop altogether the notion that ϕa says about *a* what ϕb says about *b*' with Dirichlet's notion of arbitrary function presented in Chapter 1.

$\phi_e^1 x : p$
$\phi_e^2 x : q$
$\phi_e^3 x : r$
. . .

and

$\phi_e^1 y : p$
$\phi_e^2 y : q$
$\phi_e^3 y : r$
. . .

so that the expression

$$(\phi_e^1 x \equiv \phi_e^1 y) \wedge (\phi_e^2 x \equiv \phi_e^2 y) \wedge (\phi_e^3 x \equiv \phi_e^3 y) \wedge \ldots$$

is always true, a tautology. In the case where $x \neq y$, there will be a given correlation associating some p with x and $\neg p$ with y. Then $\phi_e^1 x \equiv \phi_e^1 y$ becomes self-contradictory, and therefore (2) also becomes self-contradictory. In this way, using the words of Robert Fogelin in 'Wittgenstein on Identity', 'Ramsey's definition mirrors the formal structure of the standard definition of identity' (Fogelin 1983: 146), while avoiding any commitment to the proscribed identity of indiscernibles.

Wittgenstein knew perfectly well what was at stake.[18] These developments were conceived in a different spirit not only because Ramsey was further extensionalizing a system, the ramified theory of types of *Principia Mathematica*, which was already too extensional for Wittgenstein, but also because Ramsey's own functions in extension were inextricably emeshed with his view that mathematical equations are tautologies (they lead to an attempt at ascribing a tautological character to the Axiom of Infinity, as we shall see below).[19] This was

[18] It is a most remarkable fact that one of the annotations by Wittgenstein of Ramsey's copy of *TLP* in 1923 (i.e. while Ramsey was visiting Wittgenstein in Puchberg) is precisely the word 'identity', written in front of 6.031! (This is intriguing enough, but should not surprise anyone.) This is reported by Casimir Lewy (1967: 422). The relevant pages of Ramsey's copy are reproduced in Nedo and Ranchetti (1983: 190–1).

[19] There is to my knowledge no written record of Wittgenstein making an explicit connection between Ramsey's 'functions in extension' and his view that mathematical equations are tautologies. But he made the connection in his lectures, as witnessed by a passage in Moore's lecture notes which was not included in his article on 'Wittgenstein's Lectures in 1930–33' (*M*). The lectures notes are kept at the University Library, Cambridge; the passage occurs in catalogue number Add. 8875, item 10.7.4, p. 73.

an opinion common among logicists which annoyed Wittgenstein enormously.[20] In fact, he almost invariably advanced the same argument against that view, which involves considerations about 'surveyability' (*WVC*, p. 35). It will be discussed fully in section 8.2. That the view that mathematics are of a tautological character is a misreading of the remarks on arithmetic in the *Tractatus Logico-Philosophicus* is easily seen from section 2.1. In 1930 Wittgenstein was rejecting the assimilation of mathematical equations with tautologies in no uncertain terms, claiming that 'calculus cannot have anything to do with tautology' (*WVC*, p. 106). Waismann's *Theses* are very clear on that point. The analogy between mathematics and logic which prompted Wittgenstein to write that 'mathematics is a logical method' (*TLP*, 6.2) was the following:

> In mathematics, too, there is an operation that corresponds to the operation which generates a new sense from the senses of given propositions, namely the operation which generates a new number from given numbers. That is, *a number corresponds to a truth-function.*
>
> Logical operations are performed with propositions, arithmetical ones with numbers. The result of a logical operation is a proposition, the result of an arithmetical one is a *number.* (*WVC*, p. 218)

But this analogy has limits. This is clear in the *Tractatus Logico-Philosophicus* when, after writing that '[m]athematics is a method of logic' (6.234) Wittgenstein writes that '[t]he essential of mathematical method is working with equations' (6.2341). This much is repeated in Waismann's *Theses*: 'arithmetic considers *equations* between numbers'; and Waismann also makes it clear that Wittgenstein held the view that '*equality is not an operation*' (*WVC*, p. 218). Accordingly, equality is just the indication that 'different operations lead to the same result' and 'not the expression of an operation' (p. 218). What would count in logic as the equivalent of an arithmetical equation is not a truth-function (the only operations in logic) but a statement that the results of two operations are the same. There is no such thing (p. 218).

Ramsey's improvements upon *Principia Mathematica* were thus conceived in a spirit entirely different from that of the *Tractatus Logico-Philosophicus*, and Wittgenstein had no other choice but to attack the key notion of 'function in extension'. In a letter to Ramsey

[20] Ramsey was not alone in that view; it was shared by members of the Vienna Circle. See e.g. Hahn (1980: 25, 34).

dated June 1927, Wittgenstein raised some objections to it, to which Ramsey replied in August of the same year, in a letter to Schlick.[21] First, Wittgenstein rewrote Ramsey's (2) as $Q(x,y)$. He then remarked, rightly, that according to this definition $Q(x,y)$ comes out as a tautology whenever x and y have the same meaning and as a contradiction whenever they have a different meaning. This is exactly how the 'functions in extension' mirror Russellian identity, and therefore exactly the conclusion Wittgenstein had to oppose. His argument was therefore to the effect that 'neither "$Q(x,y)$" although it is a very interesting function, nor any propositional function whatever, can be substituted for "$x = y$" ' (*WVC*, p. 190). One must be clear about Wittgenstein's intentions here. Indeed, Ramsey implied in his letter to Schlick (and in its two drafts) that Wittgenstein has shown that '$Q(x, y)$ does not *say* that x *and* y are identical' (*WVC*, p. 191; Ramsey 1990: 342, 345). Commentators such as Fogelin have been misled by this remark into thinking that Wittgenstein misunderstood and misstated Ramsey's position (Fogelin 1983: 147–8). As I read Wittgenstein, this was not at all the point of his argument, which was rather that nothing is needed to replace the sign = and that Ramsey's device $Q(x, y)$ will run into troubles of its own. One should compare the above quotation from Wittgenstein with Ramsey's own statement: 'I only proposed $Q(x, y)$ as a substitute for the symbol $x = y$, used in general propositions and in defining classes' (*WVC*, p. 191). Wittgenstein's argument is as follows: let us take first the case when $Q(x, y)$ is said to come out as a contradiction. In this case, there is a function, say ϕ_e^1, where p is associated with a and $\neg p$ with b. This function was dubbed by Wittgenstein the 'critical' function. He carried on as follows: we know that a and b have a different meaning here, but to state the contrary in this case, i.e. that a and b have the same meaning, is not nonsensical. Because if it were nonsensical, to say 'a and b do not have the same meaning' would also be nonsensical, for '*the negation of nonsense is nonsense*' (*WVC*, p. 189). Let us now suppose, wrongly, that $a = b$, since it is not nonsensical to suppose so. We could then substitute a for b in the critical function ϕ_e^1. But then we have $\phi_e^1(a)$: p and $\phi_e^1(a) : \neg p$ as the result of the substitution. The critical function becomes therefore ambiguous, 'nonsensical'. If

[21] Both are reproduced in *WVC*, pp. 189–91. Two drafts of Ramsey's reply have been edited recently, along with some fascinating material from his posthumous writings (1991). There are further remarks on Ramsey's 'functions in extension' in Wittgenstein's later writings, to be discussed below.

the critical function is nonsensical, so is Q(a, b), which does not come out as a contradiction.

Let us now take the other case, where $c = d$, so that Q(c, d) comes out as a tautology. By analogy with the previous case, it is not nonsense to say $c \neq d$, so let us suppose that $c \neq d$. Q(c, d) remains a tautology, however, since there could be no critical function in the logical product, because 'this sign would become meaningless' (*WVC*, p. 190). Wittgenstein could then conclude:

Therefore, if '$x = y$' were a tautology or a contradiction and correctly defined by 'Q(x, y)', 'Q(a, b)' would not be contradictory, but nonsensical (as this supposition, if it were the supposition that 'a' and 'b' had the same meaning, would make the critical function nonsensical). And therefore '$\neg$ Q(a, b)' would be nonsensical too, for the negation of nonsense is nonsense.

In the case of c and d 'Q(c, d)' remains tautologous, even if c and d could be supposed to be different (for in this case a critical function cannot be supposed to exist). (*WVC*, p. 190)

He has thus reached his conclusion, which was not that one cannot substitute Q(x, y) for $x = y$, but rather that nothing is solved by doing so.

Whatever worth Wittgenstein's arguments have, it should be clear, however, that he was not guilty of the blatant misunderstanding of Ramsey's position attributed to him by Fogelin. Even if there is an argument against Wittgenstein, it is not to be found in Ramsey's reply. As I mentioned earlier, Ramsey took it that the point of Wittgenstein's argument was that Q(x, y) does not say that $x = y$. To this he agreed 'entirely'; but this was not Wittgenstein's point, which was rather that Q(x, y) runs into its own trouble and does not fare better than identity. Ramsey added, however:

it still seems to me that Q(x, y) is an adequate substitute for $x = y$ as an element in logical notation. We always use $x = y$ as part of a propositional function which is generalized, and in any such case we shall get the right sense for the resulting general proposition if we put Q(x, y) instead. (*WVC*, p. 191)

This passage and the equivalent ones in the two drafts do not contain an answer to Wittgenstein's specific objection in his letter; Ramsey was rather restating his faith in the standard interpretation as the only correct interpretation. Indeed, in the second draft Ramsey gave a few cases where he thought that by substituting Q(x, a) for $x = a$, as in substituting

$$\exists x\ F(x) \wedge Q(x, a)$$

for

$$\exists x\ F(x) \wedge (x = a)$$

(this is a variant of one of Ramsey's examples), 'we shall get the right meaning for the whole proposition' (Ramsey 1990: 346). But what did Ramsey mean by the 'right meaning' here and in the last quotation from his letter? It is of course the standard interpretation of the general proposition. Why? Because Ramsey thought that Russell's notion of identity (in terms of agreement with respect to all predicative functions) was inadequate. Recall, however, that Wittgenstein went in the opposite direction in the *Tractatus Logico-Philosophicus*, since for him

$$\exists x\ F(x) \wedge (x = a)$$

is adequately rendered by $F(a)$ (*TLP*, 5.47). As I shall argue in Chapter 4, we should rather look at systems such as the logic-free equational calculus of Wittgenstein's student Louis Goodstein as the natural development of his ideas. Provided, of course, that arithmetical equality is distinguished from and not considered as a special case of the general concept of logical identity. Again, Ramsey took the opposite stance, rejecting the view of mathematics as equations in the *Tractatus Logico-Philosophicus* as 'a ridiculously narrow view of mathematics' (Ramsey 1978: 168).

Another argument in Wittgenstein's original letter was directed at Ramsey's attempt to save the Axiom of Infinity, in the last pages of 'The Foundations of Mathematics' (Ramsey 1978: 210–12). As I have said, Russell's Axiom of Infinity may be taken to state that there is an infinity of distinguishable individuals. Now according to Ramsey, the Axiom of Infinity has, in his own system, a different meaning from Russell's: 'on my system, which admits functions in extension, the Axiom of Infinity asserts merely that there are an infinite number of individuals' (p. 210). In order to justify his view—which is contrary to Russell's and Wittgenstein's—that this is not merely a question of fact, Ramsey needed to show that the 'axiom', if it meant anything, had to be a tautology. He proceeded as follows: first, he took as an example the simpler 'There is an individual' or $\exists x\ (x = x)$ and argued that it has to be a tautology:

> Now what is this proposition? It is the logical sum of the tautologies $x = x$ for all the values of x, and is therefore a tautology. But suppose there were

no individuals, and therefore no values of x, then the above formula is absolute nonsense. So, if it means anything, it must be a tautology. (p. 211)

Secondly, once it is recognized that 'There is one individual' is a tautology, Ramsey could show, on the same grounds, that all sentences of the form:

'There is at least 2 individuals', . . . 'There are at least $\aleph_0$ individuals'

are tautologies. As for Wittgenstein's argument against this reasoning, I shall simply quote his letter:

Your mistake becomes still clearer in its consequences; viz. when you try to say, 'There is an individual'. You are aware of the fact that the supposition of there being no individual makes

$(\exists x).\ x = x$

'absolute nonsense'. But if 'E' is to say 'There is an individual', '¬ E' says: 'There is no individual'. Therefore from '¬ E' follows that 'E' is nonsense. Therefore '¬ E' must be nonsense itself, and therefore again so must be 'E'.

The case lies as before. 'E', according to your definition of the sign '=', may be a tautology right enough, but does not say, 'There is an individual'. Perhaps you will answer: Of course it does not say, 'There is an individual', but it *shows* what we really mean when we say, 'There is an individual'. But this is not shown by 'E', but simply by the legitimate use of the symbol '$(\exists x)$..', and therefore just as well (and as badly) by the expression '$\neg\ (\exists x).\ x = x$'. The same, of course, applies to your expressions, 'There are at least two individuals' and so on. (*WVC*, pp. 190–1)

In his reply, Ramsey more or less concedes the argument to Wittgenstein. This is, I think, clearer in the two drafts of the letter, where a reluctant Ramsey nevertheless voiced some objections. A more thorough study is needed here, in order to evaluate the strength of both Wittgenstein's argument and Ramsey's objections; but at any rate it seems clear—and this is the point which needs to be emphasized here—that Wittgenstein's positions and Ramsey's are irreconcilable, the former wanting to get rid of $\exists x\ (x = x)$ while the latter thought he could replace it by $\exists x\ Q(x, x)$. Ramsey put himself in a difficult situation by, on the one hand, accepting Wittgenstein's criticisms in the *Tractatus Logico-Philosophicus* and, on the other hand, wanting to reform *Principia Mathematica* in a direction opposite from Wittgenstein's. The latter's argument consists mainly in pointing out that one cannot do what Ramsey tried to achieve if one adopts the standpoint of the *Tractatus Logico-Philosophicus*. In fact, the above argument against Ramsey's attempt to save the Axiom of Infinity is a textbook answer.

The foregoing remarks show that Wittgenstein was clearly aware of Ramsey's intention to re-establish the standard interpretation by the introduction of his functions in extension, and that he rejected it as vehemently as he rejected Russell's use of the identity sign. One will find in *Philosophical Remarks* a clear statement of Wittgenstein's position in terms of the standard/non-standard distinction:

> Ramsey's theory of identity makes the mistake that would be made by someone who said that you could use a painting as a mirror as well, even if only for a single posture. If we say this, we overlook that what is essential to a mirror is precisely that you can infer from it the posture of a body in front of it, whereas in the case of the painting you have to know that the postures tally before you can construe the picture as a mirror image. (*PR*, § 121)

As Fogelin rightly points out (1983: 150), Wittgenstein used the relationship between the mirror and its images as a metaphor for the internal relation between a function and its value for different arguments: he held it as essential to a function that one can infer its value from its argument in the same way as one can infer from looking at a mirror the position of the body in front of it. This relationship must be given by a rule and it is a mistake, according to Wittgenstein, to identify a function with its extension, as Ramsey did. For him there should be no such thing as an arbitrary function in the sense of Dirichlet. So his metaphor hides a profound commitment to the non-standard interpretation, since it makes sense as a critical remark only from this viewpoint. No surprise, then, that, immediately after reproducing the same metaphor in *Philosophical Grammar*, Wittgenstein mentions Dirichlet's notion of arbitrary function: 'If Dirichlet's conception of function has a strict sense, it must be expressed in a *definition* that uses the table to define the function-signs as equivalent' (*PG*, p. 315).[22] A page later, Wittgenstein makes a connection with Ramsey's functions in extension. The latter must be some such definition and, for that reason, cannot really be functions:

> What is in question here is whether functions in extension are any use; because Ramsey's explanation of the identity sign is just such a specification

[22] Cf. *WVC*, pp. 102–3, part of which has already been quoted in Ch. 1 above, where Wittgenstein sees in Dirichlet's notion of an arbitrary function the origin of set theory. It is difficult to find remarks of a more Kroneckerian spirit.

by extension. Now what exactly is the specification of a function by its extension? Obviously, it is a group of definitions, e.g.

fa = p Def.
fb = q Def.
fc = r Def.

These definitions permit us to substitute for the known propositions 'p', 'q', 'r' the signs 'fa' 'fb' 'fc'. To say that these three definitions determine the function f(ξ) is either to say nothing, or to say the same as the three definitions say.

For the signs 'fa' 'fb' 'fc' are no more function and argument than the words 'Co(rn)', 'Co(al)' and 'Co(lt)' are. (*PG*, pp. 316–17)

To make sense of this remark one must go back to Ramsey's original definition, where a function in extension, say ϕ^1_e, correlates one arbitrary proposition with each individual. Wittgenstein's point seems to be that each function must be defined as a table:

$\phi^1_e x : p$
$\phi^1_e y : q$
$\phi^1_e z : r$
. . .

Wittgenstein's point was simply that an arbitrary correlation can only be specified by a list; therefore the members of the list $\phi^1_e x$, $\phi^1_e y$, $\phi^1_e z$, and so on, are not really of the function–argument form: no rule or law is provided. This is why he referred to F(a), F(b), and F(c) as empty 'signs'. This last move was miscontrued by Fogelin as a misunderstanding of Ramsey's 'functions in extension' (1983: 149–50), while it was simply an indication that Wittgenstein did not recognize these signs F(a), F(b), and F(c) as being of the function–argument form, i.e. that F is not a bona fide function. This point is in essence similar to Wittgenstein's objection to Russell's theory of cardinal numbers, which was discussed earlier.

Wittgenstein returned to Cambridge in 1929, where he reportedly met with Ramsey many times a week until Ramsey's untimely death in January 1930. As we shall see in the next chapter, these discussions had a profound influence on both philosophers. Ramsey's posthumous writings show without the shadow of a doubt that he converted in 1929, precisely the year of his conversations with Wittgenstein, to the finitist standpoint. Adopting a finitist point of view meant adopting a non-standard interpretation. It also meant abandoning the Axiom of Infinity. In a manuscript dated August 1929, 'Note on

Theories', Ramsey wrote: 'It is obvious that mathematics does not require the existence of an infinite number of things' (1990: 236).[23]

3.3. THE AXIOM OF CHOICE AND NUMERICAL EQUIVALENCE

Another topic where Wittgenstein's adoption of a non-standard position is to be expected is the Axiom of Choice, since the standard interpretation of higher-order quantifiers is, for our purposes, practically equivalent to the general notion of an arbitrary function, and thus equivalent to the full Axiom of Choice. To see this, one just has to think of 'dependent' quantifiers:

$$\forall x\, \exists y\, \mathrm{F}(x, y)$$

The equivalence between this formula and

$$\exists \mathrm{F}\, \forall x\, \varphi(x, \mathrm{F}(x))$$

with the standard interpretation of the existential quantifier would amount to a form of the Axiom of Choice. It is worth noticing that, amazingly enough, Wittgenstein does not account for 'dependent' quantifiers in the *Tractatus Logico-Philosophicus*.[24] There are also no remarks on the Axiom of Choice in that book and very little—in fact only one remark—in the transitional period.[25] Nevertheless, this remark is instructive, provided one carefully sets the backdrop.

[23] As Nils-Eric Sahlin pointed out, it is not clear, however, if it was because Ramsey changed his mind about general propositions (see section 4.1 for details) that he abandoned the Axiom of Infinity or the other way round (Sahlin 1990: 105).

[24] This major defect was pointed out in Fogelin (1976: ch. 6) and Soames (1983) proposes a solution. It remains true that no such solution was proposed by Wittgenstein, who seems to have remained unaware of this lacuna.

[25] There are a few more remarks on the Axiom of Choice in the later writings, which are of little interest here, except that they indicate that Wittgenstein never really accepted it. These remarks appear in the midst of a discusssion of one of Wittgenstein's slogans according to which one may be said to understand a mathematical proposition 'when one can apply' it or 'when one has a clear picture of its application' (*RFM* v, § 25). (This is actually a theme originating in the transitional period. To Schlick and Waismann, Wittgenstein said in 1931: 'I understand a proposition by applying it. Understanding is thus not a particular process; it is operating with a proposition' (*WVC*, p. 167). Thus Wittgenstein claims that the sense of a mathematical proposition is determined by its application (*PG*, pp. 310–11), a claim that does not cohere well with the assertion, seen in Ch. 1, that it is the proof which fixes the content of the mathematical proposition.) In this context, Wittgenstein refers to the Axiom of Choice as being an example of a proposition that one 'is even unable to apply' (*RFM* v,

In a simple and modern form, the Axiom of Choice says that if A is any set of non-empty, (pairwise) disjoint sets, then there exists a set C with just one member in common with each member of A. Those members or 'choice' elements are chosen arbitrarily because no rule or law for the selection can be specified. This arbitrary selection is a procedure which poses no conceptual problems in finite contexts, because it is assumed that it can be done 'in principle'. But the extension to infinite selections, denumerable or non-denumerable, caused a major controversy around the turn of the century.[26] Even today, the Axiom of Choice is still a cause of concern even to classical mathematicians who are not prone to worrying about foundational matters, and some mathematicians try to use it as sparingly as possible.[27] Ernst Zermelo was the first to use it overtly in his first proof of the Well-Ordering Theorem (Zermelo 1904), which was at the centre of a heated controversy between French mathematicians. For according

§ 25). Wittgenstein has an even stronger condemnation later on: 'We might say: if you did not understand *any* mathematical proposition better than you understand the Multiplicative Axiom, then you would *not* understand mathematics' (*RFM* vii, § 33). (As we shall see below, 'Multiplicative Axiom' was Russell's name for the Axiom of Choice.) Perhaps this is too strong a claim, since it seems obvious that mathematicians always knew how to apply the Axiom of Choice. But there is nothing wrong with the weaker claim that mathematicians did not have a 'clear picture' of the applications of the Axiom of Choice, at the time Wittgenstein wrote his remarks (1942–4), since mathematicians did not know much about which proofs needed it and which did not. (Needless to say, this is hardly the case today.) One famous example is the Schröder–Bernstein or Cantor–Bernstein theorem, which states that for two cardinal numbers κ and λ, if $\kappa \leq \lambda$ and $\kappa \geq \lambda$ then $\kappa = \lambda$. The theorem was conjectured, but not proved, by Cantor and the original proof by Dedekind made use of a principle equivalent to the Axiom of Choice. Schröder and Bernstein devised a proof, with the so-called 'mirrors', which does not need anything equivalent to the Axiom of Choice. (See Enderton 1977: 147.) In connection with this issue, it is worth pointing out an important consequence of Gödel's work on the consistency of set theory with the Axiom of Choice and the Continuum Hypothesis, in *The Consistency of the Continuum Hypothesis* (Gödel 1940), which he himself overlooked. It was first noticed by Georg Kreisel: 'A consequence of Gödel's work on the consistency of the axiom of choice and the continuum hypothesis is this: if an experimental theorem can be proved in standard set theory from these axioms it can also be proved without them: one "relativises" the proof to so-called constructible sets and classes, and observes that an arithmetical theorem is its own relativised form since the integers are absolute' (Kreisel 1956: 165). To put it briefly, Gödel's result says that if one possesses a proof in Zermelo-Fraenkel's system of set theory (ZF) of a number-theoretical statement α which uses the Axiom of Choice, then one can find a proof in ZF of α which makes no use of it. (This result does not hold, however, if one makes strong assumptions about large cardinals, outside ZF.) Thus one sees that uses of the Axiom of Choice in number theory or algebraic theories such as Galois's theory are in principle eliminable.

[26] For a detailed history of the Axiom of Choice, see Moore (1982).

[27] For an example of this cautious attitude, see Garling (1986: pp. vii, 14).

to Borel, Baire, and Lebesgue all admissible objects must be defined 'in a finite number of words', and from this viewpoint an uncountably infinite, arbitrary selection of the kind needed if the set under consideration is the continuum, as it was in Zermelo's proof, was simply not conceivable. In a direct comment on Zermelo's choice function, Émile Borel insisted that 'it would be necessary to give at least a theoretical means of determining the [choice] element' in each subset, and he simply commented that fulfilling this requirement seemed 'extremely difficult' in the case of the continuum (Borel 1905: 194). Actually, while Borel thought that reasonings with uncountably many arbitrary choices were 'outside the domain of mathematics' (1905: 195), he left open the possibility of countably or denumerably many arbitrary choices, thus accepting the Axiom of Countable Choice, which will be introduced in a moment. Incidentally, it should be said that Borel accepted denumerably many choices for reasons linked with probability theory (to be discussed briefly in section 7.2), and that it was he who first introduced the notion of 'choice sequences'—the term even having its origin in the dispute on the Axiom of Choice.[28]

In the famous exchange of letters on this topic mentioned above in section 3.1, Baire and Lebesgue went further than Borel (while Jacques Hadamard sided with Zermelo). Lebesgue simply thought that countable choice was faring no better than the full Axiom of Choice: 'I see sometimes difficulties as grave in reasoning with a denumerable infinite number of choices as in reasoning with a transfinite number of them' (Borel 1914: 156). In fact, Lebesgue thought that the underlying question was that of mathematical existence: 'can we prove the existence of a mathematical entity without defining it?' His standpoint, which he admitted to be 'close to that of Kronecker and Drach', was that 'we can only build solidly if *we admit that we can prove the existence of an entity only by defining it*' (Borel 1914: 154). In his letter, Baire also went further than Borel in admitting only finitely many choices (Borel 1914: 151–2). For Baire too we must limit ourselves to definable entities and 'despite appearances, everything must be brought back to the finite' (Borel 1914: 152).

This typically Kroneckerian stance was also adopted by the Norwegian Thoralf Skolem, who developed in the early 1920s a system of primitive recursive arithmetic not without affinities with some

[28] See Troelstra (1982: 466).

ideas of Wittgenstein (as we shall see in section 4.1). He described his work as 'a consistently finitist one; it is built upon Kronecker's principle that a mathematical definition is a genuine definition if and only if it leads to the goal by means of a *finite* number of trials' (Skolem 1967b: 333). Skolem could accept neither the reduction of arithmetical notions to set-theoretical ones nor the Axiom of Choice:

So long as we are on purely axiomatic ground there is, of course, nothing special to be remarked concerning the principle of choice . . . but if many mathematicians—indeed, I believe, most of them—do not want to accept the principle of choice, it is because they do not have an axiomatic conception of set theory at all. They think of sets as given by specification of arbitrary collections; but then they also demand that every set be definable. We can, after all, ask: What does it mean for a set to exist if it can perhaps never be defined? It seems clear that this existence can be only a manner of speaking, which can lead only to purely formal propositions—perhaps made up of very beautiful *words*—about objects *called* sets. But most mathematicians want mathematics to deal, ultimately, with performable computing operations and not to consist of formal propositions about objects called this or that. (Skolem 1967a: 300)

Not all reactions to Zermelo's Axiom of Choice were negative, of course. Ernst Steinitz's much vaunted result on existence and uniqueness of an algebraic closure for a given field,[29] which is a cornerstone of 'modern' algebra (as developed in Emmy Noether's school), makes one of the first striking uses of the Axiom of Choice. Steinitz himself added a note of both caution (in the name of *Methodenreinheit*) and optimism about the fate of Axiom of Choice:

Many mathematicians are still opposed to the axiom of choice. With the increasing recognition *that there are questions in mathematics which cannot be decided without this axiom*, the opposition to it must increasingly disappear. On the other hand, it seems expedient in the interest of the purity of method to avoid the said axiom when the nature of the question does not require its application. I have endeavoured to make these limits conspicuous. (Steinitz 1930: 170–1)

[29] See Steinitz (1930: 33) and the editorial n. 43 by Baer and Hasse (Steinitz 1930: 153). Bart van der Waerden's *Modern Algebra* contains a description of Steinitz's existence and uniqueness theorems for countable fields (van der Waerden 1931: 193–6). I hasten to add that in Kronecker's approach to algebraic number theory, which is in many ways more interesting than the 'modern' approach derived from Dedekind, there is simply no need to discuss uniqueness (Weyl 1940: 12).

This is the background against which Wittgenstein's remark on the Axiom of Choice is best seen. Closer to home (Cambridge), he knew of the Axiom under its Russellian name of Multiplicative Axiom.[30] Russell's ontological conviction forced him to introduce all functions in a non-extensional manner. We saw in sections 2.3 and 3.2 that Ramsey did not share his ontological conviction. He also criticized Russell's interpretation of the Axiom of Choice in 'The Foundations of Arithmetic', because the class whose existence was asserted had for them to be definable by a propositional function, and his claim was that this rendered the Axiom doubtful. Instead, Ramsey proposed that we understand by class 'any set of things homogeneous in type not necessarily definable by a function which is not merely a function in extension' (1978: 208–9) and he did not ask that the choice function be definable in formal terms. This had the advantage of rendering the Axiom a 'most evident tautology' (1978: 209). Moving in the opposite direction to Ramsey, Wittgenstein wrote in 1929:

> What gives the multiplicative axiom its plausibility? Surely that in the case of a finite class of classes we can in fact make a selection (choice). But what about the case of infinitely many subclasses? It's obvious that in such a case I can only know the law for making a selection.
>
> Now I can make something like a random selection from a finite class of classes. But is that conceivable in the case of an infinite class of classes? It seems to me to be nonsense. (*WA* i, p. 90; *PR*, § 146)

This passage indicates clearly that, although Wittgenstein had qualms about a finite number of selections, he simply rejected in the strongest possible terms the idea of an infinite number of selections for which no rule could be stated. This reaction clearly puts Wittgenstein in the camp of Kroneckerian finitists such as Baire, Borel, Lebesgue, and Skolem. Hermann Weyl, a qualified mathematician if ever there was one, used almost the same words in *The Continuum*: 'the notion that an infinite set is a "gathering" brought together by infinitely many individual arbitrary acts of selection, assembled and then surveyed as a whole by consciousness, is nonsensical' (Weyl 1987: 23). One must

[30] In *PM* the Axiom of Choice is introduced in *88: 'If κ is a class of mutually exclusive classes, no one of which is null, there is at least one class μ which takes one and only one member from each member of κ' (*PM* i, p. 536). Russell probably gave it the name 'Multiplicative Axiom' because he first came across it in connection with cardinal multiplication, i.e. while having to construct a class for the product of a denumerable infinity of cardinals. Russell's famous example, which will be presented below, is that of the millionaire who has bought $\aleph_0$ pairs of boots and $\aleph_0$ pairs of socks and wishes to pair them. (Incidentally, this example is not a case of infinite multiplication.)

distinguish a strongly finitist viewpoint, which would even close the door on weaker versions of the Axiom, from a weaker finitist viewpoint, which accommodates such weaker versions, although for both viewpoints the full Axiom of Choice remains unacceptable. (Baire, Lebesgue, and Skolem can be said here to adopt a strong finitist stance, and this was already seen as insufficient by Borel.) In fact, intuitionists adopt the Axiom of Countable Choice, also known in the literature as AC-NN or AC_{00}:

$$\forall x\, \exists y\, \varphi(x, y) \rightarrow \exists F\, \forall x\, \varphi(x, F(x))$$

with the variable F ranging over functions from **N** to **N**.[31] This weaker form is acceptable to intuitionists because of their interpretation of the quantifiers. As Dummett points out in *Elements of Intuitionism*, on an intuitionist interpretation the antecedent states that for every x one possesses an effective procedure that yields a y for which $\varphi(x, y)$. This is only made explicit by the consequent, where the function F is constructive and gives y when applied to x (Dummett 1977: 52–3).[32] Wittgenstein does state in his solitary remark that the idea of an infinite number of selections is nonsense *only if no rule could be stated*. We can thus infer—and that is about all that we can infer—that the idea of an infinite selection is perfectly acceptable to him if, in his own terminology, a law for making the selections is known. This is precisely the reasoning behind a constructive version of the Axiom of Choice such as that adopted by intuitionists.[33]

As we saw in section 3.1, Wittgenstein, by doing away with identity, was also getting rid of the theory of cardinal numbers in *Principia Mathematica*. In his conversations with Schlick and Waismann in the early 1930s, he put forward a new argument directed against the notion of numerical equivalence. This argument was directed not just towards *Principia Mathematica* but principally towards Frege's use of the relation of equivalence or one-to-one correspondence in his definition of cardinal numbers. The most striking difference between Frege's and Dedekind's definition of natural numbers is that the former associated, at the time of writing *Foundations of Arithmetic*, numbers with extensions of concepts and

[31] On this and other weaker versions of the Axiom of Choice acceptable to intuitionists, see Troelstra and van Dalen (1988: 189–90).

[32] See also Troesltra and van Dalen (1988: 189 f.).

[33] Lello Frascolla is responsible for having opened my eyes to this point.

had consequently to provide a definition of cardinal numbers instead of a definition of ordinal numbers. The key principle of Frege's analysis in *Foundations of Arithmetic* is introduced in § 63:

The number of Fs = the number of Gs if and only if F ≈ G

where F ≈ G says that F and G are in one-to-one correspondence, i.e. they are numerically equivalent.[34] So Frege's principle says that two sets have the same number if and only if they are numerically equivalent.[35] Frege's analysis can only hold if he can show that the equivalence relation ≈ is somewhat conceptually prior to the very notion of number. But he only offered in support of his claim an argument from analogy according to which we can attain the concept of direction only after having grasped the relation of parallelism and not the other way round (Frege 1980a: §§ 64–7). We should thus see that we can attain the concept of sameness of number only after having grasped the prior relation of equivalence (one-to-one correspondence) and not the other way round. As I take it, Wittgenstein's strategy was simply to argue against Frege by denying that the equivalence relation ≈ is indeed conceptually prior to the notion of number: were the latter to be prior to the former, Frege's analysis of numerical identity would involve a circularity. To Schlick and Waismann, Wittgenstein presented his argument in his usual informal manner:

Imagine I have a dozen cups. Now I wish to tell you that I have got just as many spoons. How can I do it?

If I had wanted to say that I alloted one spoon to each cup, I would not have expressed what I meant by saying that I have just as many spoons as cups. Thus it will be better for me to say, I can allot the spoons to the cups. What does the word 'Can' mean here? If I meant it in the physical sense, that is to say, if I mean that I have the physical strength to distribute the spoons

[34] This is the principle known since George Boolos's work as 'Hume's Principle'. See Boolos (1987; 1990). It has been recently shown by Crispin Wright (1983: 154–69) and George Boolos, in the previously mentioned papers and in Boolos (1988), that Hume's Principle implies the infinity of natural numbers (Peano's second axiom). This rediscovery of Frege's result has sparked a justly deserved renewal of interest in Frege's philosophy of mathematics. See the papers in Demopoulos (1995).

[35] Frege is then able to define the cardinal number of a class C as the class of all classes in one-to-one correspondence to C (Frege 1980a: § 72). He thus succeeds in defining numbers *without* using numerical terms, i.e. without circularity. He could also prove that each class has a unique cardinal number and that the cardinals of classes in one-to-one correspondence are identical. With 0 being defined as the number of the empty class (Frege 1980a: § 74), and with the help of the relation of immediate succession (which is defined in § 76), Frege could then define the number 1 as the class of all classes numerically equivalent to 0 (§ 78), and so on . . .

among the cups—then you would tell me, We already knew that you were able to do that. What I mean is obviously this: I can allot the spoons to the cups because there is the right number of spoons. But to explain this I must presuppose the concept of number. It is not the case that a correlation defines number; rather, number makes a correlation possible. This is why you cannot explain number by means of correlation (equinumerosity). You must not explain number by means of correlation; you can explain it by means of possible correlation, and this precisely presupposes number. You cannot rest the concept of number upon correlation. (*WVC*, pp. 164–5)

and he went on sharpening his argument:

When Frege and Russell attempt to define number through correlation, the following has to be said:

A correlation only obtains if it has been *produced*. Frege thought that if two sets have equally many members, then there is already a correlation too . . . Nothing of the sort! A correlation is there only when I actually correlate the sets, i.e. as soon as I specify a definitive relation. But if in this whole chain of reasoning the *possibility* of correlation is meant, then it presupposes precisely the concept of number. Thus there is nothing at all to be gained by the attempt to base number on correlation. (p. 165)

He also wrote in the *Philosophical Remarks*:

Can I know there are as many apples as pears on this plate, without knowing how many? And what is meant by not knowing how many? And how can I find out how many? Surely by counting. It is obvious that you can discover that there are the same number by correlation, without counting the classes.

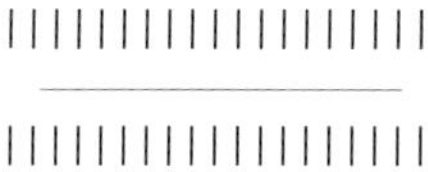

In Russell's theory only an *actual* correlation can show the 'similarity' between the classes. Not the *possibility* of correlation, for this consists precisely in numerical equality. Indeed, the possibility must be an *internal* relation between the extensions of the concepts, but this internal relation is only given through the equality of the 2 numbers. (*PR*, § 118)

If I read him properly, Wittgenstein claimed first that in order to specify what is meant by saying that two sets are numerically equal, one should not say that there *is*, but rather that there *could be* a one-to-one correspondence between the two sets. His objection was that one cannot explain this 'could' without reference to numbers, i.e. without any circularity, since one can only say that there are just as many Fs and Gs because there 'could' be a one-to-one correspondence, and there 'could' be such a one-to-one correspondence only as far as the number of Fs and Gs is concerned.[36] Before establishing whether or not Wittgenstein's objection really hit its intended target, namely Frege's use of equivalence, it is worth pointing out that this very argument was utilized (and popularized by the same token) by Friedrich Waismann in his *Introduction to Mathematical Thinking* (1951: 107–13).[37] Michael Dummett, who knew of Wittgenstein's argument through its use by Waismann, has dismissed it on two occasions, most recently in his book *Frege: Philosophy of Mathematics* (1991: 148–9).[38] It should be said also that Louis Goodstein, one of Wittgenstein's students in the early 1930s, also used the same argument. In *Constructive Formalism*, Goodstein wrote:

> How do we know whether two elements correspond or not? A cup standing on a saucer have an obvious correspondence, so too a bird carrying a twig in his beak, or a pencil mark on a sheet of paper; but it is patently false to say there *is* a correspondence between the members of any two classes (of the same number of terms); and if one says that, even when there is no actual correspondence, such a correspondence always *can* be established, the possibility to which we refer must be a logical possibility, a consequence of, not a condition for, the two classes having an equal number of terms.
>
> *The concepts of number and function are defined by the transformation rules for number and function signs. It is not a one-to-one related pair of classes that determines a function, but a function which determines a pair of one-to-one related classes.* (Goodstein 1951: 19)[39]

[36] Wittgenstein would later describe this idea in a more colourful way by saying of Russell's definition of numerical equivalence (which is similar to Frege's): 'He puts the cart before the horse' (*PG*, p. 355).

[37] There is an objection somewhat similar to Wittgenstein's in Hermann Weyl's *Philosophy of Mathematics and Natural Sciences*, in the midst of an argument to the effect that the concept of ordinal number, and not cardinal number, is the primary one (Weyl 1959: 34–5). But the notion of time is involved in Weyl's argument, and we know at least that Wittgenstein did not agree with it nor with Weyl's conclusion on the primacy of ordinal numbers. See *WVC*, p. 84. For a useful discussion, see Bouveresse (1992: 226–8).

[38] Dummett had previously criticized Waismann's argument in his 1963 paper, 'Realism' (Dummett 1978a: 150–1).

[39] Goodstein had no use for numerical equivalence for the simple reason that his own equational calculus, of which his book *Constructive Formalism* contains a presentation, is not set-theoretical. Of all things, there are no quantifiers, and presumably no way to express the Axiom of Choice. The affinities between Wittgenstein's logical ideas in the transitional years and Goodstein's equational calculus will be discussed in section 4.2.

One has first to notice that Wittgenstein emphasized the 'possibility' versus the 'actuality' of the one-to-one correspondence and we may dub his argument, following Dummett, the 'modal argument' (Dummett 1991: 148). Dummett pointed out, however, that 'Frege invokes no modal notions: his definition is in terms of there being a suitable mapping' (pp. 148–9). Wittgenstein's objection should be reformulated thus: Frege must provide us with a criterion for the existence of relations which can be framed without circularity. We are facing once again the issue of the interpretation of second-order quantification. The issue lies at the heart of Wittgenstein's philosophy of mathematics, and it thus constitutes a crucial test for the cogency of his anti-Platonist standpoint.

The reasoning behind the idea of numerical equivalence is very simple.[40] When presented with (finite) sets of Fs and Gs and asked 'Are there as many Fs as Gs?', one simply has, say, to use one's pen and join them one by one. This way, one can construct a relation. The procedure is unquestionably sound in strictly finite cases, but problems occur when the number of Fs or Gs is so large that the procedure is practically impossible. (Problems of this nature were raised by Wittgenstein; they will be discussed in Chapter 8.) In the infinite case, Russell's example (presumably known to Wittgenstein) of the millionaire who bought $\aleph_0$ pairs of boots and $\aleph_0$ pairs of socks serves perfectly to illustrate the kind of problem one faces (Russell 1919: 126). The answer to the question 'How many pairs of boots and pairs of socks had the millionaire?' is actually $\aleph_0$. But in order to arrive at this result one is faced with the problem of selecting one of each pair of socks. In the case of boots one can distinguish between left and right and thus select all the right boots and then all the left boots. This cannot be the case with socks, since there are presumably no distinguishing properties. The Platonist will simply assume the full Axiom of Choice in order to allow for selection of the socks and thus establish the needed pairing. Thus, according to Dummett, the Axiom of Choice tells us 'when there are enough members of each two sets . . . for there to be a one-one map of either on to the other; and it does so without circularity' (Dummett 1991: 149). Dummett's point here is that, assuming the Axiom of Choice allows for the comparability of cardinals, for any transfinite cardinal, one can tell with help of the Axiom of Choice whether it is greater than or equal to

[40] See e.g. Enderton (1977: 128–9).

$\aleph_0$. So, although the standard interpretation remains questionable, it need not involve a circularity, and Wittgenstein's argument misfires.

It is difficult, however, to fault Wittgenstein for not having thought carefully about Choice, since it is at any rate not clear prima facie whether a weaker version of the Axiom of Choice which would be acceptable to Wittgenstein, according to the above argument, is of itself sufficient to ensure comparability of cardinals. At any rate, the fact that Wittgenstein's objection is not convincing for those who adopt Frege's stance does not of itself vitiate Wittgenstein's own non-standard approach. (It ought to be stressed that the very fact that Wittgenstein felt the need to put forth this objection shows, again, his commitment to a non-standard approach; his modal argument simply would not make sense otherwise.) Moreover, Wittgenstein's remarks should be seen as being more than an unsuccessful attempt at raising the spectre of circularity. One could distinguish two levels in his argument: first, the recognition that the only acceptable claim is that there are just as many Fs and Gs because there *could* be a one-to-one correspondence and, secondly, the objection of circularity, which we can put aside since it would not be recognized by a Fregean. Wittgenstein's claim that we could only say 'there are just as many Fs and Gs' because there *could* be and not because *there already is* a one-to-one correspondence brings to the fore an underlying assumption made by the Platonist, namely that the correspondence already obtains before it has been carried out. To paraphrase Dummett, the correspondence appears to give us information about a state of affairs whose existence is independent of our carrying it out (Dummett 1978a: 151). One finds the perfect example of this Platonist attitude in a passage of the *Grundgesetze der Arithmetik*, where Frege asked: What happens when we correlate two concepts? His answer was that the situation is similar to that of drawing an auxiliary line in geometry, because the act of drawing does not 'create' the line:

> What then are we really doing if we (correlate) for the purpose of proof? Evidently, something similar to when we draw an auxiliary line in geometry. Euclid, whose method can in many cases serve as an example of rigour, has for this purpose his postulates, which allow one to draw certain lines. But to draw a line may no more be seen as an act of creation than may the determination of a point of intersection. On the contrary, in both cases we only bring into consciousness, only grasp, what was already there. For the

proof, it is essential only that there is thus something there. (Frege 1966: § 66)[41]

Wittgenstein's own viewpoint is to the effect that there is nothing—no one-to-one correspondence—existing before one has effectively correlated the Fs to the Gs. Only after there has been a correlation can we say that there a numerical equivalence between the two given sets obtains. This is the fundamental anti-Platonist idea which Wittgenstein expressed later on by saying that 'the proof doesn't *explore* the essence of the two figures, but it does express what I am going to count as belonging to the essence of the figures from now on.—I deposit what belongs to the essence among the paradigms of language' (*RFM* i, § 32). (We see here that Wittgenstein's attitude towards numerical equivalence is supported by his view that it is the proof which fixes the content of a mathematical proposition, which will be discussed briefly in section 6.1.) It was after quoting by heart those words of Frege that Wittgenstein continued the conversation with Schlick and Waismann, adding:

This dictum sounds very paradoxical. It is connected with Frege's distinction between 'objective' and 'real'.

What Frege means is evidently that it is possible to draw a line. But possibility is not yet reality. A straight line is drawn only when it has been drawn. And this is how it is with numbers too. (*WVC*, p. 165)

Therefore, although Wittgenstein was perhaps not able to provide a convincing argument against the notion of numerical equivalence,[42] he detected with characteristic flair an important Platonist assumption which is even recognized by Dummett as standing 'in need of philosophical justification' (Dummett 1978a: 151), namely the assumption that the correspondence between the Fs and the Gs already obtains before the correlation has been carried out.

[41] I have here used an unpublished translation of pt. ii of Frege's *Grundgesetze der Arithmetik* by Jason Stanley and Richard Heck.

[42] Unlike many topics which ceased to interest him after the mid-1930s, numerical equivalence takes a special place in Wittgenstein's later writings and lectures. See e.g. Lecture XVI in *LFM*, pp. 152–61, and *RFM* i, § 25–40.

4

Quantification and Finitism

4.1. WEYL, HILBERT, AND RAMSEY

In March 1928 Brouwer gave two lectures in Vienna, 'Mathematik, Wissenschaft und Sprache' and 'Die Struktur des Kontinuum' (Brouwer 1975: 417–28, 429–40). We know that Wittgenstein attended the first.[1] The exact nature of Wittgenstein's reaction to Brouwer's lecture is a matter of controversy. Eyewitness accounts by Herbert Feigl and Karl Menger[2] lend support to the idea that it marked the return of Wittgenstein to philosophy, his excitement after the talk contrasting markedly with his lack of interest during previous meetings with members of the Vienna Circle. Taking their lead, among other things, from this anecdote, some commentators went so far as to see in Brouwer's intuitionism the source of Wittgenstein's later philosophy.[3] But a number of commentators have recently taken the opposite view. For example, Peter Hacker points out that Wittgenstein's 'excitement after the lecture may just as well have been a reaction to Brouwer's *misconceptions*' (Hacker 1986: 120). One of the aims of this study is to show that Wittgenstein's philosophy of mathematics has much less in common with intuitionism than is usually assumed, although some strategical moves are barely distinguishable.[4] At least one such move cannot be overlooked because of its importance for the development of Wittgenstein's philosophy: it concerns quantification theory. We shall see in the next section that Wittgenstein's stance on quantification was, however, even more radical than that of intuitionists. This is a topic about which Wittgenstein changed his mind around 1929. One can see this from

[1] There is no reason to believe, contrary to the claim made in van Peursen (1970: 96), that Wittgenstein attended both lectures.

[2] Feigl's account is reported in Pitcher (1964: 8 n.), while Menger's is found in the more recent Menger (1994: ch. 10).

[3] e.g. Hacker (1972: 98–104).

[4] Thus I agree with Hacker's revised assessment of the relations between Wittgenstein and Brouwer in Hacker (1986).

G. E. Moore's lectures notes: 'in the *Tractatus* [Wittgenstein] had made the mistake of supposing that an infinite series was a logical product—that it *could* be enumerated, though we were unable to enumerate it' (*M*, p. 298). Georg Henrik von Wright also reported that in one of their first conversations in 1939, Wittgenstein told him that 'the biggest mistake he had made in the *Tractatus* was that he had identified general propositions with infinite conjunctions or disjunctions of singular propositions' (von Wright 1971: 123).

The search for the source of Wittgenstein's new ideas on quantification leads directly to Hermann Weyl.[5] In his paper 'Über die neue Grundlagenkrise der Mathematik' (1968: ii. 143–80) Weyl was officially abandoning his earlier predicativist position of *The Continuum* (Weyl 1987) and siding with Brouwer, whose ideas on a host of topics he set forth to explain. In doing so, he actually developed original ideas on topics such as quantification.[6] According to him, propositions containing an existential quantifier ranging over the natural numbers do not possess the full status of judgement (*Urteil*) or, as I would prefer to say, statement:[7] for F, a decidable predicate, the proposition $\exists x\ F(x)$ can be asserted only if one knows already a specific number a of which one can show that it satisfies the predicate F, that is one abstracts $\exists x\ F(x)$ from $F(a)$, or

$$F(a) \rightarrow \exists x\ F(x)$$

For this reason, Weyl described existential propositions as *Urteilabstrakte* or 'judgement-abstracts':

> *An existential sentence*—e.g. 'there is an even number'—*is not at all a judgement which asserts a fact in the proper sense.* Existential facts are an empty invention of logicians. '2 is an even number': this is a real judgement expressing a fact; 'there is an even number' is only a *judgement-abstract* obtained from this judgement. (Weyl 1968: ii. 156)

[5] Georg Henrik von Wright has previously noted the parallels between Weyl (and Ramsey) and Wittgenstein, to be described in this chapter, in a lengthy footnote to *Truth, Knowledge, and Modality* (von Wright 1984: 109–10), and Ulrich Majer (1988: 551) has also hinted at a connection between Weyl and Wittgenstein, via Ramsey. See also Sahlin (1995: 152–4).

[6] To my knowledge, the claim that Weyl's ideas on quantification do not square with those of Brouwer was first made by Majer (1988).

[7] In this chapter and the following one, I shall use 'proposition' to mean something which is put forward for consideration but which remains unasserted. A 'statement' will then be understood as an asserted proposition, and the expressions 'assertion' and 'statement' will be used interchangeably.

Likewise, propositions containing a universal quantifier ranging over the natural numbers are not conceived of as statements by Weyl, who called them *Anweisungen auf Urteile* or 'instructions', 'requests', or 'rules for judging': 'The general "Every number has the property F"—e.g. "for each number m, $m + 1 = 1 + m$"—is equally little a real judgement, it is rather a general *rule for judging*' (p. 157). Weyl used the following example: if three pieces of chalk are exhibited to me and I say, 'All the pieces of chalk are white', then this proposition is a statement which is either true or false, since it is an abbreviation for a finite conjunction

'This piece is white' & 'This piece is white' & 'This piece is white'

This is the assertion of a proper statement because the pieces of chalk are exhibited and an enumeration is possible. But if the domain is infinite, as in the case of universally quantified arithmetical propositions, the exhibition becomes impossible and the proposition cannot be interpreted as an abbreviation for an infinite disjunction. Such a proposition, Weyl would say, has 'obviously no meaning' (Weyl 1968: ii. 152).[8]

An elementary example such as Euclid's theorem on the infinity of prime numbers may help us understand Weyl's idea. The theorem says that 'prime numbers are more than any assigned multitude of prime numbers'. Its classical proof works by *reductio ad absurdum*: one begins by supposing that there is a greatest prime number, p_n. We should therefore be able to list all the prime numbers:

$$p_1 \ldots p_n$$

We then define the number N:

$$N = [p_1 \times p_2 \times p_3 \times \ldots \times p_n] + 1$$

This number N is either prime or composite. If it is prime, then we have a contradiction, since it would be bigger than all the prime numbers smaller or equal to p_n and there would be more than n prime numbers. If it is composite, it must be divisible exactly by a prime number. But this prime divisor cannot be any prime smaller or equal

[8] It is worth noticing that Weyl thought that anyone insisting, as he thought Brouwer did, on the distinction between the finite and the infinite was not hitting the right spot. Indeed, according to Weyl, when elements—say, the pieces of chalk—are in finite number but cannot be exhibited one by one, i.e. if the domain is finite but, for some reason or another, 'unsurveyable', the general proposition is not a genuine statement (Weyl 1968: iii. p. 151).

to p_n because they would all leave a remainder of 1. Therefore there must be another prime number, bigger than p_n. From Weyl's viewpoint, the proposition 'there exists a prime number x such that $n < x \leq N$' expresses a proper statement, because with N one remains within a finite domain. Moreover, the proof gives us enough information to find the next prime number. But if the domain is infinite as in the case of the proposition 'there exists a prime number x such that $n < x$', the exhibition is impossible and the proposition cannot be interpreted as an abbreviation for an infinite disjunction:

$$n + 1 \text{ is prime} \vee n + 2 \text{ is prime} \vee n + 3 \text{ is prime} \vee n + 4 \text{ is prime} \vee \ldots$$

and unless the speaker knows already one such specific number $x > n$ of which she can show that it is prime, she is not in a position to assert $\exists x\, F(x)$, since it would be an unjustified claim. As Wittgenstein himself would say: 'Surely that is nonsense. For "If we only search long enough" has no meaning. (That goes for existence proofs in general.)' (*PG*, p. 384).

What Weyl meant when he said that an arithmetical proposition is a 'rule for judging' was that, while it is not asserting anything by itself, it justifies the deduction of an infinity of singular propositions $F(a)$, or

$$\forall x\, F(x) \rightarrow F(a)$$

In a later paper, 'The Ghost of Modality', Weyl described it as 'a *hypothetical* proposition, saying . . . only . . .: "In case you come across a certain number (whatever this number may be) you may be sure it has the property (F)" ' (1968: iii. 701). To put it another way, Weyl provided an account of propositions containing an existential or a universal quantifier ranging over the natural numbers in terms of *claims*, as opposed to *statements*. Indeed, a proposition involving unrestricted universal quantification is construed by Weyl not as a true statement but as a claim that the speaker possesses an effective means of establishing the truth of any given statement from a given infinite range. In 'The Source of the Concept of Truth', Dummett describes this distinction in the following terms:

even if we operate with the classical conception of a large class of utterances that constitutes assertions of statements with determinate truth-conditions, we shall still need to acknowledge that not all informative utterances belong to that class. Other informative utterances may be classified, not as assertions of statements, but as expressions of claims: statements are to be assessed as true or false, but claims as justified or unjustified. (1990: 4)

Dummett's example of a class of informative utterances which should be classified as claims is precisely that of arithmetical propositions involving unrestricted quantification. In this paper he attributes this point to Hilbert; we have just seen that it had already been made by Weyl. Why is it exactly that existential and universal propositions cannot be said to be genuine statements? Because, to use Dummett's words, 'the condition for someone to be entitled to make any such utterance is inseparably connected with his own cognitive position' (1990: 2). The fact that a mathematician does not possess an instance F(a), and that she is therefore not in a position to assert $\exists x$ F(x), does not entitle her to assert the negation $\neg$ F(a) of the free-variable statement, even less to assert $\forall x \neg$ F(x), and vice versa. Thus the most important consequence of this reading of general propositions is that the Law of Excluded Middle cannot be applied to them. Indeed, since the fact that a mathematician does not possess an instance F(a) does not entitle her to assert the negation $\neg$ F(a)—and this is tantamount to the assertion $\forall x \neg$ F(x)—the equivalence between $\forall x \neg$ F(x) and $\neg \exists x$ F(x) does not hold and the Law of Excluded Middle does not apply. As Weyl would say, 'it is completely senseless to negate such sentences, therefore the possibility of formulating the law of excluded middle in regard to these sentences disappears' (1968: ii. 158).

Weyl's account of general arithmetical propositions was influential. He had been a student of David Hilbert and when he changed his allegiance in 'Über die neue Grundlagenkrise der Mathematik' and enthusiastically wrote that: 'Brouwer—das ist die Revolution!'(Weyl 1968: ii. 158), Hilbert was reportedly infuriated. He began publishing a now famous series of papers on the foundations of mathematics in which he renewed his formalist standpoint, in the hope of checking the growing popularity of Brouwer's iconoclastic ideas. This is the beginning of the *Grundlagenstreit* of the 1920s and 1930s. Hilbert adopted Weyl's reading of the *alle* and the *es gibt*, for example in 'Die logischen Grundlagen der Mathematik', where he described the universal quantifier as a logical product and the existential quantifier as a logical sum, adding that with infinite domains the negation of a general judgement (*allgemeinen Urteil*) or an existential judgement (*Existentialurteil*) has no precise content (*keinen präzisen Inhalt*) (Hilbert 1935: 182).[9] In his

[9] Cf. this with the earlier view, in Hilbert (1935: 173–4).

1925 paper, 'On the Infinite', Hilbert introduced not a dichotomy but a trichotomy between 'real', 'finitary general', and 'ideal' propositions (1967b: 380–1).[10] Elementary or 'real' propositions are simply equations involving primitive recursive functions and numerals, such as

$$509483 + 675943 = 1185428,$$
$$2 + (3 + 4) = (2 + 3) + 4$$

These propositions have an intuitive content and are verifiable by direct computation. They can be negated, so the Law of Contradiction and the Law of Excluded Middle hold for them. But besides these unproblematic propositions one will encounter other finitary propositions of problematic character. These are the 'finitary general propositions' of the form:

$$x + (y + 1) = (x + y) + 1$$

This proposition being the assertion of

$$a + (b + 1) = (a + b) + 1$$

for all numerals a and b. To use our earlier example, Hilbert claimed that the proposition 'there exists a prime number x such that $n < x \leq \mathrm{N}$' is an abbreviation for a logical sum or disjunction (1967b: 377–8). Such disjunctions were admissible because with N one never leaves the domain of finite totalities. In the infinite case (unbounded general arithmetical propositions) negation is impossible because

> the proposition cannot be interpreted as a combination, formed by means of 'and', of infinitely many numerical equations, but only as a hypothetical judgement that comes to assert something when a numeral is given. (Hilbert 1967b: 378)

It is worth noticing here the strong similarity between Hilbert's notion of 'hypothetical judgement' and Weyl's notion of a 'rule for judging'. Hilbert's claim was that when the domain of quantification is infinite, propositions of the form $\forall x\, \mathrm{F}(x)$ cannot be understood as infinite logical products, and therefore cannot be negated.[11] This was

[10] This trichotomy was pointed out to me by Daniel Isaacson, in conversation. It was also presented by C. Smorynski in his paper 'Hilbert's Programme' (Smorynski 1988: 59–64). In what follows I have used Smorynski's presentation, changing only his example, and used his terminology.

[11] See also later statements by Hilbert and Bernays in the first volume of *Grundlagen der Mathematik* on *Existenzsätze* as *Partialurteile*, on the *hypothetischen Sinn* of the *allgemeines Urteil*, and about the impossibility of their negation (Hilbert and Bernays 1934: 32–4).

the very basis of Hilbert's limited rejection of the Law of the Excluded Middle.

For reasons of simplicity, in what follows I shall call the account of general arithmetical propositions shared by Hilbert and Weyl, i.e. the account of them as expression of claims as opposed to statements, the 'finitist account'. There is an important consequence of this finitist account, drawn by Hilbert, which must be emphasized: for him the sentential operators had to be truth-functional, so they could intelligibly apply only to propositions with determinate truth-conditions. Since Hilbert recognized that only justification conditions and not determinate truth-conditions can be associated with general propositions involving unrestricted quantification, he thought that not only the Law of Excluded Middle but also *the whole calculus of truth-functions* fails to apply to infinitary propositions. This is what he meant when he wrote: 'those laws that man has always used since he began to think, the very ones that Aristotle taught, do not hold' (Hilbert 1967b: 379). We shall see that this point is of central importance for the interpretation of Wittgenstein's philosophy of mathematics.

Although Hilbert agreed with the finitist critique of quantification, he did not leave the matter there. His intention was to introduce ideal (meaningless) propositions in order to restore the laws of logic: 'we must here *adjoin the ideal propositions to the finitary ones* in order to maintain the formally simple rules of ordinary Aristotelian logic' (Hilbert 1967b: 379). So Hilbert recognized the existence of transfinite formulas as ideal propositions to be added to the finitary propositions, singular and general, in order to re-establish the validity of the laws of classical logic. Such ideal propositions would contain quantifiers and would look like this:

$$\forall x \forall y\ x + (y + 1) = (x + y) + 1$$

(To be more precise, this proposition would assert associativity for infinitary constructions involving Hilbert's ε-function.) Hilbert's metamathematical programme consisted in trying to obtain a real proposition which would be the proof of the consistency of the parts of mathematics using ideal propositions. (Consistency itself had to be nothing less than a finitary general proposition.) Therefore, the problem of finitary general propositions was transformed in Hilbert's programme into the problem of a finitary consistency proof of the system resulting from the addition of these ideal propositions. The fate of

Hilbert's programme need not detain us here. Although his account of general arithmetical propositions was close to Hilbert's, Wittgenstein rejected, in parallel with Brouwer, this metamathematical programme.[12]

Moving into Wittgenstein's narrow circle of interlocutors, another author who was deeply influenced by Weyl's paper was Frank Ramsey. The orthodox image of Frank Ramsey is that of the ultra-Platonistic philosopher of the 1926 paper 'The Foundations of Mathematics', which I presented in sections 2.3 and 3.2. But his posthumous papers indicate that he moved towards finitism during the last two years of his life (1928–9),[13] changing his views about quantification by adopting Weyl's account of general propositions as 'founts of judgements, or rules for judging' (Ramsey 1990: 235).[14] Ramsey's change of mind was set in motion by his reading, in 1928, Norman Campbell's *Physics. The Elements* (1919). Campbell distinguished two layers within a scientific theory: the first layer consists of a group of propositions about 'some collection of ideas which are characteristic of the theory' that are called 'hypotheses'; the second layer, called the 'dictionary', consists of a group of propositions relating these hypothetical ideas to 'some other ideas of a different nature', i.e. relating the hypotheses to verifiable statements. In Campbell's words, 'it must be possible to determine, apart from all knowledge of the theory, whether certain propositions involving these (latter) ideas are true or false' (Campbell 1919: 122). In a manuscript dated from 1928—therefore before Wittgenstein's return to Cambridge—Ramsey presented Campbell's standpoint, which

[12] Wittgenstein's objections to Hilbert's metamathematical programme are various. He objected mainly that '*calculation with letters is not a theory*' (*WVC*, p. 136) and therefore that no piece of mathematics could solve philosophical problems. Against Hilbert's emphasis on the need for a proof of consistency, he argued that hidden contradictions ought not to be feared because they are harmless until discovered and can be avoided, once discovered, by ad hoc stipulations. On this last point, see Gerrard (1987: ch. 4) and Wrigley (1986).

[13] This change of mind was noticed at the time of Ramsey's death by Russell, who said of the 'Last Papers' published by Richard Braithwaite in 1931 that they 'show a tendency towards the views of Brouwer' (Russell 1931: 481), and by Braithwaite himself, who wrote that by 1929 Ramsey 'was converted to the finitist view which rejects the existence of any actual infinite aggregate' (Braithwaite 1931: p. xii). Alice Ambrose also mentioned this change of mind in her paper 'Finitism in Mathematics' (1935: 188, 340). Nevertheless, it has been largely overlooked since, and the prevailing view of Ramsey remains, inaccurately, that of the arch-Platonist of 1926.

[14] For a discussion of the relations between Ramsey and Weyl, see Majer (1989; 1991).

he associated with that of Hertz (Ramsey 1991: 35, n. 1), in these terms:

> There is . . . a more radical philosophy of science . . . on which science begins with observations and laws which assert observed uniformities; and these laws are then explained by theories which introduced undefined entities and relations. Some of the statements which a theory makes about these undefined entities are to be interpreted by means of a 'dictionary' in such a way that they can be proved true or false by observation. But other statements about undefined entities have no such interpretation, and are regarded as having no 'truth', except such as can be derived from the satisfactoriness of the theory of which they form part. (p. 33)

About propositions which are not interpreted by means of a 'dictionary', Ramsey added this comment: 'These so-called propositions do not therefore express judgements . . . they are interesting as showing that a large body of sentences, which appear to express judgements and are manipulated according to the laws of formal logic may not express judgments at all' (p. 34).

One will not find in Campbell's book a statement clearly supporting Ramsey's claim that 'hypotheses' that are not related by help of the 'dictionary' to verifiable statements do 'not express judgments at all'.[15] Ramsey may have transformed Campbell's notions as a result of taking in ideas from Hermann Weyl's 1921 paper, 'Über die neue Grundlagenkrise der Mathematik', the content of which he had already known for some time.[16] In 'General Propositions and Causality', he expressed the view that there are two kinds of general proposition: if the domain is finite and objects are all concretely given, then these are simply equivalent to conjunctions of (singular) statements, but when the domain is open-ended, as in the case of propositions such as 'all men are mortal', general propositions are what he called 'variable hypotheticals'. Like Weyl's *Urteilsanweisungen*, these variable hypotheticals do not possess the full status of statements. Their status was described by Ramsey in terms strikingly similar to Weyl's: 'Variable hypotheticals are not judgements but rules for judging "If I meet a ϕ, I shall regard it as a ψ".

[15] On Campbell and Ramsey, see Sahlin (1990: 134).

[16] As Sahlin pointed out, Ramsey had known Weyl's paper 'Über die neue Grundlagenkrise der Mathematik' for some time, since he describes Weyl's *Urteilsabstrakte* and *Urteilsanweisungen* in 'Mathematical Logic' (Ramsey 1978: 218). (The relevant passages of Weyl's paper are reproduced by Ramsey in item 007-04-01 of the Ramsey Archives, Hillman Library, University of Pittsburgh.) But he seemed to have remained unimpressed by Weyl's ideas until 1928–9. See Sahlin (1995: 152–4).

This cannot be *negated* but it can be *disagreed* with by one who does not adopt it' (Ramsey 1978: 137). For Ramsey, too, general arithmetical propositions were not genuine statements but, rather, variable hypotheticals. In support, I shall simply give one quotation, from another 1929 manuscript, 'The Formal Structure of Intuitionist Mathematics':

> It seems then that a general mathematical proposition does not correspond to a judgement in the way a singular proposition does, although by substitution it leads to such judgements and truth-functions of any finite number of such judgements. When we have proved such a proposition we can, of course, make the judgement that we have proved it (and the judgement that any instance of it is true) but this is not an equivalent to the proposition itself e.g. 'I have not proved *p*' is not the same as 'I have proved not-*p*'. (Ramsey 1990: 204)

Before continuing, one final remark on Weyl's 1921 paper. Its real interest lies in his account of dependent quantifiers:

$$\forall x\, \exists y\, \mathrm{F}(x, y)$$

It was Weyl's account of these which interested Ramsey so much in 1929. It has already been pointed out (section 3.3) that there is no account of dependent quantifiers in the *Tractatus Logico-Philosophicus*. Ramsey adopted Weyl's account in his paper 'Principles of Finitist Mathematics', where he claimed that 'the proper method seems to be Weyl's' (Ramsey 1990: 256). Weyl's account has two stages. First, one constructs a law (*Gesetz*), that is a function

$$\mathrm{F}(x) = y$$

which generates out of every number x a new number y. In this way one can introduce the general proposition

$$\forall x\, \varphi(x, \mathrm{F}(x))$$

Secondly, from every singular sentence derived from this general proposition, one can abstract an existential sentence of the form

$$\exists x\, \varphi(a, x)$$

The real meaning of the *es gibt* comes therefore from the introduction of the function $\mathrm{F}(x) = y$, from which one derives the existence of the individuals y. This is why Weyl said that the *es gibt* must include the *alle* and not vice versa (Weyl 1968: ii. 159).[17]

[17] This is the 'method' with which Ramsey agreed in 1929. On this point, see Majer (1989: 247 f.).

Ramsey also extended his finitist account of general propositions to causal laws: 'when we assert a causal law we are asserting not a fact, not an infinite conjunction, nor a connection of universals, but a variable hypothetical which is not strictly a proposition at all, but a formula from which we derive propositions' (1978: 147).[18] In Ramsey's account of the laws of physics, Campbell's notion of 'hypothesis' seems to have been combined with Weyl's claim that there are no infinite conjunctions, to produce an altogether new notion of physical law, one that differs from Campbell's notion of theory since for the latter it is the hypotheses of the theories that are not provable or disprovable, not the particular laws which they explain. Here, even these laws are not assessable in terms of truth and falsity.

4.2. THE NEW LOGIC OF 1929

On 18 January 1929 Wittgenstein returned to Cambridge, after an absence of more than fifteen years. He had numerous encounters with Ramsey—they reportedly met many times a week during term—from that moment until his stay in Vienna, during the Christmas break.[19] (Ramsey died in January 1930, before Wittgenstein's return from Vienna.) It is difficult to believe that such frequent discussions did not play an important role in the evolution of each man's thoughts dur-

[18] Ramsey's conception of causal laws was to lead to a conception of scientific theories which is akin to intuitionism and, contrary to widespread belief, the exact opposite of the view popularized by Carnap as being his. This is clearly explained in ch. 6 of Sahlin's *The Philosophy of F. P. Ramsey* (1990). See also Majer (1989; 1991).

[19] There are many references in Wittgenstein's manuscripts to these 'vigorous' exchanges on logic, as he described them (*WA* i, p. 4). One also finds allusions to these conversations in Ramsey's posthumous writings, such as the remarks contained in 004-21-04 of the Ramsey Archives, Hillman Library, University of Pittsburgh. Moreover, it seems that Wittgenstein left his MS 106 in Ramsey's hands, since we find excerpts in item 004-23-01, also in the Pittsburgh Archives. The remarks copied down by Ramsey are contained in *WA* i, pp. 53–81. It is not clear to my mind who influenced whom during the conversations in 1929. In 'Ramsey and Wittgenstein on Scientific Theories', Rosaria Egidi (1991) holds that Ramsey's change of mind was the result of Wittgenstein's influence. But there are reasons to believe that Ramsey was not influenced by Wittgenstein: he knew Weyl's papers quite well (see n. 16 above) and his thoughts had already started to shift before Wittgenstein's return to Cambridge, as a result of reading Campbell's book. On Ramsey and Wittgenstein, I strongly recommend Sahlin's insightful paper 'On the Philosophical Relations between Ramsey and Wittgenstein' (1995), which deals in particular with probability, a topic that I shall not take up in this book.

ing that year; a year during which they both experienced fundamental changes of mind. In the preface of the *Philosophical Investigations*, Wittgenstein acknowledged in no uncertain terms his debt to Ramsey:

> For since begining to occupy myself with philosophy again, sixteen years ago, I have been forced to recognize grave mistakes in what I wrote in that first book. I was helped to realize these mistakes—to a degree which I myself am hardly able to estimate—by the criticism which my ideas encountered from Frank Ramsey, with whom I discussed them in innumerable conversations during the last two years of his life. (*PI*, p. viii)[20]

There is evidence that Wittgenstein and Ramsey discussed infinity and quantification,[21] and Moore's lecture notes, already quoted at the beginning of the previous section, clearly show that Wittgenstein realized that his idea that quantifiers are to be understood as infinite conjunctions or disjunctions was mistaken (*M*, pp. 297 f.).[22] In one particular passage of Moore's notes, we find Wittgenstein speaking of the class of primary colours as 'defined by grammar' (*M*, p. 297). Here an enumeration is possible, and the proposition 'in this square there is one of the primary colours' is identical with the logical sum: 'in this square there is either red or blue or yellow'. Wittgenstein then claimed that his mistake in the *Tractatus Logico-Philosophicus* was to suppose that in all classes defined by grammar general propositions were identical with a logical product or a logical sum, as in the case of primary colours. What were his reasons to see a mistake in his earlier conception? Because he saw, from 1929 onwards, an infinitely long conjunction (or disjunction) as an impossibility, not because of human limitations but because 'infinite' is not a number. It is, as Wittgenstein put it, a 'possibility in the symbolism' (*LWL*, p. 17). Therefore, Wittgenstein could only see the universal quantifier as a logical product when the dots following the expression 'and so on'

[20] It is now generally admitted that Wittgenstein's claim that he had had conversations with Ramsey 'during the last two years of his life' is an exaggeration, since when Wittgenstein arrived in Cambridge in January 1929 he had not seen Ramsey since 1925, and since Ramsey died in January 1930. Therefore, they could have met only for a year, which seems to have been ample time for fruitful exchanges.

[21] In a letter from Ramsey to Trinity College cited in *M*, p. 254. See Wrigley (1995: 166).

[22] See also Alice Ambrose's notes from the same lectures, Waismann's notes (*WVC*, p. 45), and later remarks in *Philosophical Grammar* where Wittgenstein admits that his former view is correct 'for *one* use of words "all" and "some" ', i.e. in cases such as 'In this square there is one of the primary colours' (*PG*, p. 268). This change of mind was witnessed by Russell in a report to Trinity College (Russell 1968: 200).

were, as he called them, 'dots of laziness', i.e. an abbreviation for a finite enumeration. Moore reported:

He said he had been misled by the fact that $(x) . fx$ can be replaced by $fa . fb . fc \ldots$ having failed to see that the latter expression is not always a logical product: that it is only a logical product if the dots are what he called 'the dots of laziness', as where we represent the alphabet by 'A, B, C . . .', and therefore the whole expression can be replaced by an enumeration; but that it is not a logical product where, *e.g.* we represent the cardinal numbers by 1, 2, 3, . . . where the dots are not the 'dots of laziness' and the whole expression can not be replaced by an enumeration. (*M*, p. 298)[23]

This reasoning bears strong similarities to the finitist account espoused by Weyl and Ramsey,[24] i.e. the account of them as expression of claims as opposed to statements. But during a meeting with Schlick and Waismann in January 1930, Wittgenstein claimed that 'Weyl lumps several different things together' (*WVC*, p. 82), and rejected Weyl's account:

Weyl pretends there may indeed be universal statements (*All-Aussage*) but they do not have negations, on the ground that an existential statement (*Existenzaussage*) is a 'judgement-abstraction' (*Urteilsabstrakt*) and that only construction (finding a number) tells us anything. But in reality these are two completely different things—a universal statement (*All-Aussage*) is correctly expressed by means of induction and as such it naturally cannot be negated. (*WVC*, p. 81)

[23] See also *AWL*, p. 6.

[24] Diamond (1991) develops an interpretation of Wittgenstein as a 'realist' philosopher. In a seemingly Wittgensteinian spirit, she disqualifies by the same token Ramsey's philosophy, as it is displayed in his 1929 paper 'General Proposition and Causality', as 'unrealistic'. She begins by presenting a conception of realism as she finds it outside philosophy, in particular 'in connection with novels and stories' (p. 40), where realism is said to be characterized in particular by a 'certain kind of attention to reality: to detail and particularity' and by the fact that in such novels 'certain things do not happen', i.e. that for this form of realism 'fantasy' is excluded. Diamond further believes that Wittgenstein was referring precisely to this literary notion of realism when he described empiricism as a mistaken attempt at realism, in this passage, dating from 1943–4: 'Not empiricism and yet realism in philosophy, that is the hardest thing. (Against Ramsey.)' (*RFM* vi, § 23). Since she also claims that 'Ramsey defended an empiricist view of causality, a view closely related to Hume's' (p. 42), Diamond's belief that Wittgenstein was opting for her literary notion of realism is supported in part by the fact that Wittgenstein was criticizing Ramsey in this very passage. I cannot go into all the details here of Diamond's reading of Ramsey's 1929 paper and of her attempt at showing that it is based on 'fantasy', hence the '*un*realism' (Diamond 1991: 65). I have objected (Marion 1994) that Diamond misreads both philosophers, and the content of this chapter, with its numerous textual references, should provide ample evidence in favour of my reading.

Wittgenstein's account of Weyl's argument is slightly erroneous. He took it that a general proposition such as $\forall x\ F(x)$ cannot have a negation because such a negation would be equivalent to a purely existential proposition, although the latter can only be a judgement-abstract. But Weyl was not exactly 'lumping' universal and existential propositions together: he thought that negations of general propositions with unrestricted quantification are not *umfangs-definit* and this is why these cannot be negated (Weyl 1968: ii. 155). The above quotation is nevertheless instructive. The fact that Wittgenstein misunderstood Weyl's position is a good indication that he could not have been (at least directly) influenced by him. (In fact, we saw in section 2.1 that for Wittgenstein an operation does not assert anything; only its result does (*TLP*, 5.25); I shall show in Chapter 5 that his position in the early 1930s can be seen to be a development from the *Tractatus Logico-Philosophicus*.) Moreover, one learns that Wittgenstein thought that his account of general propositions differed from that of Weyl precisely because of his own view that they are 'correctly expressed by means of induction'. In the same meeting with Schlick and Waismann, Wittgenstein reiterated this point: 'An assertion (*Aussage*) about *all* numbers is not represented by means of a statement (*Satz*), but by means of an induction. Induction however, cannot be denied, nor can you affirm it, for it does not assert anything' (*WVC*, p. 82). There are two claims here: first, there is the claim that generality is expressed correctly not by a statement of the form $\forall x\ F(x)$ but by an 'induction'. Secondly, since it is claimed that such 'inductions' are not proper statements, they cannot be negated. This is in line with the finitist account of general propositions involving unrestricted quantification as the expression of claims as opposed to statements, the consequence of which being that the whole calculus of truth-functions does not apply—such propositions are neither true nor false[25]— since the sentential operators, being truth-functional, could apply only to propositions with determinate truth-conditions.

[25] When presented with this consequence in Wittgenstein's lectures, Moore voiced his disagreement: 'Wittgenstein has not succeeded in removing the "uncomfortable feeling" which it gives me to be told that "3 + 3 = 6" and "$(p \rightarrow q\ \&\ p)$ entails q" are neither true nor false' (*M*, p. 289). It is hoped that the explanations given in this chapter will help removing this 'uncomfortable feeling'. Juliet Floyd (1994) takes the opposite approach and argues that Wittgenstein did not answer Moore's objection because he did not need to.

I shall come back to these two points. I must first explain Wittgenstein's account of proofs by induction (he also used the expression 'recursive proofs'), of which these points are consequences. During the early 1930s Wittgenstein held the view that there are two kinds of proof in mathematics: algebraic proofs and proofs by induction (*M*, p. 301; *WVC*, p. 135). According to Wittgenstein, the distinguishing feature of proofs by induction was that the proposition proved does not appear as the last step of the proof, so that the proof is not one of the proposition *per se*; it has instead the form of a 'template' for particular proofs. (In the following I use the expression 'template' as a synonym of the Wittgenstein's expression 'proof by induction', and related terms.) Here are two telling quotations:

> A recursive proof is only a general guide to an arbitrary special proof. A signpost that shows every proposition of a particular form a particular way home. (*PR*, § 164)
>
> Of course the so-called 'recursive definition' isn't a definition in the customary sense of the word, because it isn't an equation, since the equation 'a + (b + 1) = (a + b) + 1' is only part of it. Nor is it a logical product of equations. Instead, it is a law for the construction of equations. (*PG*, p. 432)

This account of proofs by induction is very similar to what Hilbert and Bernays had to say in their classic book *Grundlagen der Mathematik* about Skolem's primitive recursive arithmetic.[26] We can thus begin to see the deep connection between Wittgenstein's ideas and Skolem's free-variable finitism. In 1923 Skolem published his famous paper, 'Begründung der elementaren Arithmetik durch rekurriende Denkweise ohne Anwendung scheinbarer Verändlichen mit unendlichen Ausdehnungsbereich' (Skolem 1967b), of which Wittgenstein is said to have owned a copy.[27] Skolem had just read *Principia Mathematica* and, as a finitist solution to the problem of the paradoxes of set theory, he proposed to develop a fair part of elementary arithmetic without the use of unbounded quantifiers (Skolem was also avoiding the theory of types): 'If we consider the general theorems of arithmetic to be functional assertions and take the recursive

[26] See Hilbert (1967b: 380) and, for details, Hilbert and Bernays (1934: 298–9). Cf. Gandy's discussion of 'proofs by example' (1982: 139): he claims that 'our faith in, and understanding of, primitive recursive arithmetic rests on our recognition that its theorems may be given proofs by example', and points out that this is true of proofs by induction or of their equivalent in Goodstein's equational calculus.

[27] See the editor's footnote to *PR*, § 163.

mode of thought as a basis, then that science can be founded in a rigorous way without the use of Russell and Whitehead's notions "always" and "sometimes"' (Skolem 1967b: 304). What Skolem called in this passage the 'recursive mode of thought' consisted of the use of primitive recursive definitions for the introduction of new functions and of mathematical induction for proofs. Skolem allowed the use of bounded quantifiers only as shorthand notation. He used free variables only when they ranged over infinite domains. In the resulting system of primitive recursive arithmetic, generality is expressed by the use of a free variable. This is remarkably similar to Wittgenstein's conception of generality in the *Tractatus Logico-Philosophicus* (see section 2.2). Similarities extend also to the account of proofs by induction: in a free-variable system of arithmetic such as Skolem's, a proof of the free-variable formula F can be seen as a 'template' or 'schema' yielding proofs of the instances of F, by simple transformations. Consider an instance of F resulting from the replacement of the functions of F by their values, that is a variable-free numerical formula: the proof of F can be transformed into a proof of this instance by replacing the variables by the numerals and computing the function values. (Here the rule of mathematical induction is replaced by a sequence of conditionals.) The resemblance with Wittgenstein's conceptions, as seen in the quotations above, is striking, and it justifies to a large extent labelling Wittgenstein as a 'finitist'.[28]

In order to extract the philosophical content of the notion of template, Wittgenstein asked: 'To what extent, now, can we call such a guide to proofs the proof of a general proposition?' (*PR*, § 164), and his answer was, in an application of the distinction between 'saying' and 'showing' (*TLP*, 4.121–4.1212), that the template does not *assert* its generality, but *shows* it, in that it 'allows us to see an infinite possibility':

[28] William Tait has defended the thesis that 'finitist reasoning is essentially primitive recursive reasoning in the sense of Skolem' (Tait 1981: 524), referring precisely to Skolem's 1923 paper and his 'recursive mode of thought'. I am aware that Tait's claim, originally set forth in Tait (1968), has been criticized by Kreisel (1970), who defends the view already expressed in Kreisel (1958b) that finitist functions are not just primitive recursive functions but all those which can be proved in first-order number theory. There is no need to arbitrate this debate here; it suffices to say that Tait's characterization of 'finitism', for whatever it is worth, fits Wittgenstein's conceptions very well, as they are presented here—hence the applicability of the label 'finitism' to Wittgenstein.

Its generality doesn't lie in itself, but in the possibility of its correct application. And for that it has to keep on having recourse to the induction.

That is, it does not assert its generality, it does not express it; the generality is, rather, shown in the formal relation to the substitution, which proves to be a term of the inductive series. (*PR*, § 168)

To get to the root of Wittgenstein's ideas, one must go back to one of the most important insights in the *Tractatus Logico-Philosophicus*, namely that logical constants do not refer (*TLP*, 4.0312): if no prior acquaintance with a logical object is to be presupposed by the understanding of any proposition, the conditions of the meaning must come from the constituents of the proposition themselves. This is what Wittgenstein meant in the *Notebooks* when he said that 'the proposition represents the situation—as it were off its own bat' (*NB*, p. 26). The situation is similar with respect to infinity in mathematics. Wittgenstein was asking that the 'infinite possibility' be easily read from the symbolism itself:

To explain the infinite possibility, it must be sufficient to point out the features of the sign which lead us to assume this infinite possibility, or better: what is actually present in the sign must be sufficient. . . . so everything must be already contained in the sign '$|1, x, x + 1|$'—the expression for the rule of formation. In introducing infinite possibility, I musn't reintroduce a mythical element into grammar. (*PR*, app. 1, p. 314). [29]

So, although generality is to be found in templates, they do not express it. Generality is inexpressible; it is only shown in the template by the possibility of replacing the variables by the numerals and of thus transforming the proof of F into a particular proof.

When Wittgenstein said after 1929 that templates are not genuine statements, he was not saying anything new, since he already described mathematical equations as pseudo-statements in the *Tractatus Logico-Philosophicus*:[30] mathematical equations cannot be negated and they are thus lacking bipolarity (the ability to be true *and* to be false). In *Philosophical Remarks*, Wittgenstein was still holding that mathematical equations cannot be negated in the ordinary sense of the word, because we cannot imagine under which circumstances we would be prepared to assert the negation $\neg$ F(a) of a free-variable formula F. Compare the following passage:

[29] See also *WVC*, p. 34.

[30] There are also reasons to believe that the later Wittgenstein carried on questioning the applicability of the concept of truth to mathematical sentences. See Floyd (1991).

It seems clear that negation means something different in arithmetic from what it means in the rest of language. If I say 7 is not divisible by 3, then I can't even make a picture of this, I can't imagine how it would be if 7 were divisible by 3. All this follows naturally from the fact that mathematical equations aren't a kind of statement. (*PR*, § 200)

This is not psychologism. Wittgenstein himself insisted in *Philosophical Investigations* that '"I can't imagine the opposite of this" does not mean: my powers of imagination are unequal to the task. These words are a defence against something whose form makes it look like an empirical proposition, but which is really a grammatical one' (*PI*, § 251). In our context, it is worth pointing out for those who find the expression 'I can't imagine the opposite of this' too psychologistic that both Skolem's primitive recursive arithmetic and Goodstein's equational calculus, by expressing generality by the use of a free variable, rule out the negation of unrestricted general propositions. These are not expressible in the language. This is what Wittgenstein meant when he wrote, for example, that one 'cannot deny the *generality* of general arithmetical propositions' (*PR*, § 169). In a lecture given in 1990 at the University of Oxford, 'Experience and Mathematical Necessity', Hilary Putnam claimed that if we cannot imagine under which circumstances we would be prepared to assert the negation $\neg$ F(a), we cannot attach a clear *sense* to 'F(a) might be false'. This claim was not meant by Putnam to be a 'metaphysical' thesis about the nature of certain propositions. To speak another jargon, this is not to say that F(a) is unrevisable, but that we cannot make sense of the question: 'Can F(a) be revised?' The question has no sense. Wittgenstein would say: 'Where you can't look for an answer, you can't ask either' (*PR*, § 149; *PG*, p. 377).

It should be noted that in the passage from the *Philosophical Remarks* quoted above, Wittgenstein was pointing out not that templates but that singular equations such as $25 \times 25 = 625$ could not be negated in the ordinary sense of the word, while according to the finitist account (Weyl, Hilbert) the negation of singular or 'real' propositions is *unproblematisch*. There are other passages where the same idea is expressed. For example, in his account of Wittgenstein's lectures, Moore objects to a similar example, namely that $3 + 3 = 6$, which is said by Wittgenstein to be neither true nor false (*M*, pp. 285–9). From this one may infer that Wittgenstein is also making the much stronger claim that for a general proof of F, the singular proposition F(a) would still not be informative or that an equation such as $25 \times 25 = 625$ is *sinnlos*. This radical view seems at first blush to be hardly defensible. At any rate, this view amounts to little more than the one expressed earlier in the *Tractatus Logico-Philosophicus*, and I believe that Wittgenstein was in fact at the time in the process of changing his mind. As we shall see, he was struggling to deny that such elementary equations were devoid of 'sense'. He even introduced and discussed the idea of an analogy between mathematical equations and ordinary statements which is based on the fact that equations have some 'sense'. I shall come back to this delicate issue in sections 6.1 and 6.2.

Wittgenstein drew two important consequences from his account of proofs by induction. First, since a proof by induction does not assert anything, it is not assessable as true or false. Hence the inapplicability of the Law of Excluded Middle: 'Therefore, where there is an assertion (*Aussage*) it can be negated; and where a certain structure (*Gebilde*) cannot be negated, there is no assertion (*Aussage*) either. The law of excluded middle however does not apply—simply because we are not dealing with propositions (*Sätze*) here' (*WVC*, p. 82). The parallel with the finitist account of Weyl and Ramsey is striking. It is worth pointing out in this context the following fascinating passage from Ramsey's manuscript on 'The Formal Structure of Intuitionist Mathematics':

> We cannot therefore assume that mathematical propositions in general can be made arguments to truth-functions and treated by the propositional calculus, but must examine this question afresh. It is complicated by the fact that our previous criticisms have left us without a clear conception of the nature and purpose of mathematics. Our old conception that the 'propositions' of mathematics expressed each a true judgement has been destroyed and we have as yet nothing to put in its place. (Ramsey 1990: 205)

There are passages where my interpretation provides, at this stage, a simple reading of otherwise bizarre claims. These are the passages where Wittgenstein wrote that the Fundamental Theorem of Algebra and Fermat's Last Theorem (supposing that one would have a proof, which is now the case) are not genuine statements:

> If the proof that every equation has a root is a recursive proof then that means the Fundamental Theorem of Algebra is not a genuine mathematical *proposition*. (*PR*, § 168)

> I say: the so-called 'Fermat's Last Theorem' isn't a proposition. (Not even in the sense of a proposition of arithmetic.) Rather, it corresponds to a proof by induction. (*PR*, § 189)

Wittgenstein's language is likely to mislead readers into thinking that he believed, for whatever reason, that these theorems are somewhat not really theorems of mathematics. Surely the Fundamental Theorem of Algebra is part of mathematics: there are many different proofs of it.[31] It would have been sheer foolishness on the part of Wittgenstein to deny any mathematical status to that theorem! Such prima-facie bizarre claims make perfect sense under my interpretation: since Wittgenstein thought that the result of a proof by induction is not a genuine statement (*Aussage*), because the induction does not assert anything, he would not give to the theorems thus proved the full status of statements. He was rather saying that no such theorems should be construed as statements assessable in terms of truth and falsity to which the calculus of truth-functions applies.

The second consequence is that a statement containing the universal quantifier—$\forall x\ F(x)$ understood classically—'adds something to the proof', i.e. that 'it does not follow from it' but stands proxy for it (*WVC*, p. 135). It is usually assumed that there is in the end no difference between the utterance of the free-variable form $F(a)$, supposing one has a proof of it, and the utterance of its universal closure $\forall x\ F(x)$; but Wittgenstein was, it seems to me, inclined to think that the use of the quantifier blurs the distinction that he was at pains to make. Therefore, he denied that a proof by induction, while being a proof of generality, is a proof that the property holds for 'all' numbers: 'A proof by induction, if it were a proof, would be a proof of generality, not a proof of a certain property of all numbers' (*PR*, § 168). It was for him inappropriate to speak of a proof that a certain property holds for 'all' natural numbers as if it is a statement about an infinite extension: 'But you can't talk about all numbers, because there is no such thing as all numbers' (*PR*, §124).[32]

A natural solution is to get rid of the quantifiers. In his Cambridge lectures (Lent term, 1930) Wittgenstein made the following comment:

> The rule for infinity can be expressed symbolically as follows: $[f(1), f(\xi), f(\xi + 1)]$. Note that we have to go on step by step, starting from $f(1)$. This is not the kind of generality represented by $(x)\phi x$. (*LWL*, p. 14)

He wrote something similar in the *Philosophical Grammar*:

[31] Incidentally, Goodstein has obtained, quite late in his career, a finitist version of Gauss's second proof of the Fundamental Theorem of Algebra (Goodstein 1969).

[32] See also *PR*, §126.

The point of our formulation is of course that the concept 'all numbers' is given by a structure like '| 1, ξ, ξ + 1 |'. The generality is *set out* in the symbolism by this structure and cannot be *described* by an $(x).fx$. (*PG*, p. 432)

It is clear from these quotations that Wittgenstein wanted to draw a distinction between the universal quantifier $\forall x\, F(x)$ and expressions of the type [f(1), f(ξ), f(ξ + 1)]. If we follow Wittgenstein correctly, the quantifiers $\forall x\, F(x)$ and $\exists x\, F(x)$ would be introduced only for finite sequences, because they express external or 'accidental' generality, which is acceptable only for finite extensions.[33] Perhaps he thought that limiting the use of the quantifiers to finite sequences and using expressions such as [f(1), f(ξ), f(ξ + 1)] or | 1, ξ, ξ + 1 | for adequate description of infinite series would fulfil the role of a syntactical distinction.[34] Both expressions are derived from the 'variable' [*a*, *x*, *O'x*] of the *Tractatus Logico-Philosophicus* (*TLP*, 5.2522). One should recall here from section 2.1 that this 'variable' is the general term of a series which consists only of iterations of *O* with respect to *a*:

a, *O'a*, *O'O'a*, *O'O'O'a*, *O'O'O'O'a* . . .

There is an obvious analogy with the scheme of pure iteration:[35]

$$F(a, 0)$$
$$F(a, Sx) = BF(a, x)$$

where the function $F(a, x)$ is simply the result of applying x times the function B to a. (It is as if the square brackets in Wittgenstein's

[33] Wittgenstein's finitism was quite transparent to his contemporaries. Russell had already noticed it in his introduction to *TLP* (p. 21). See also Russell (1936: 143). In *The Logical Syntax of Language*, Wittgenstein's apparent restriction to finite quantifiers, with unrestricted generality expressed by free variables, is embodied in Carnap's Language I, with its 'limited universal operator' and 'limited existential operator'—those were defined as, respectively, a finite sum and a finite product (Carnap 1937: 20–1). Carnap's discussion of Wittgenstein as an ally of intuitionism assumes that he accepted only this finitist Language I (Carnap 1937: 46–9).

[34] This account nicely fits Wittgenstein's criticisms of the Russellian symbolism in *WVC* (p. 228). A different solution would be the introduction of a third and new quantifier, ranging over infinite series, while the universal quantifier is used for finite sequences. So far as I know, only Yvon Gauthier has taken up this idea, without previous knowledge of these remarks. See Gauthier (1985) for his 'effinite' quantifier and the corresponding logical system. Gauthier has more recently noticed the parallel between Wittgenstein's remarks and his own 'effinite' quantifier (Gauthier 1991: 84). It is unlikely, however, that Wittgenstein had in mind a new quantifier for infinite sequences.

[35] This analogy has been noticed in Sundholm (1992).

expression play the role of the iteration-functional.) Raphael Robinson showed long ago that with a suitable choice of initial functions, all the primitive recursive functions can be obtained from this scheme of pure iteration (Robinson 1947). This, again, is a strong indication of the affinity between Wittgenstein's ideas and quantifier-free arithmetic. Wittgenstein was quite serious about using such expressions. For example, when Skolem defines addition by recursion in his 1923 paper:

$$a + 0 = a$$
$$a + (b + 1) = (a + b) + 1$$

Wittgenstein rewrote the second step thus:

$$\begin{array}{lcl} a + (1 + 1) & = & (a + 1) + 1 \\ \downarrow & & \downarrow \\ a + (\xi + 1) & & (a + \xi) + 1 \\ a + ((\xi + 1) + 1) & & ((a + \xi) + 1) + 1 \end{array}$$

with the 'rule' ranging 'over the series $|1, \xi, \xi + 1|$' (*PG*, p. 433). His remarks about Skolem's primitive recursive arithmetic (Skolem 1967b) make only sense within this context.[36] Indeed, he made a

[36] Stuart Shanker (1987: 199) claims that Wittgenstein's 'sustained attack . . . on Skolem's definition of "recursive proof" must be discomfiting for those critics who wish to portray Wittgenstein as a revisionist at heart . . . After all, given Wittgenstein's hostility to the classical interpretation of the quantifiers, together with Skolem's declared Kroneckerian/finitist objectives, one would have expected that if his interest really did lie in the direction of strict finitism Wittgenstein would have greeted Skolem's results with enthusiasm. But quite to the contrary, Wittgenstein regarded Skolem's proof as a paradigmatic example of the type of confusion which so often results when mathematicians attempt to "translate the calculus into the signs of word-language".' These remarks deserve some comments. If one conveniently forgets the confusion between finitism and strict finitism, Shanker's claim is, plainly, that the gist of Wittgenstein's remarks on Skolem's definition of 'recursive proof', presumably through his discussion of the proof of the law of associativity, shows that Wittgenstein was critical of finitism—this claim being diametrically opposed to mine. To begin with, the fact that someone is critical of some aspects of Skolem's 1923 paper does not imply that that person is not a finitist. At any rate, I do not see any 'sustained attack' in Wittgenstein's remarks on Skolem: it will be easily granted that Skolem did not avoid transgressing Wittgenstein's idiosyncratic saying/showing distinction, which he probably never knew. But the fact that Wittgenstein wanted to interpret recursive proofs in terms of the latter does not imply that he rejected Skolem's work as a 'paradigmatic example' of 'confusion'. It was shown above that the usual interpretation of Skolem's recursive proof is very close to Wittgenstein's. I think that Wittgenstein understood very well the difference between generality in, say, *Principia Mathematica* and in Skolem's arithmetic—the analyses in this section will show this conclusively—and that Shanker plainly misreads both Wittgenstein and Skolem when he adds in support of his claims: 'the brunt of [Wittgenstein's] offensive was centred on Skolem's

number of remarks on Skolem's proof by induction of the law of associativity,

$$a + (b + c) = (a + b) + c$$

in *Philosophical Remarks* and *Philosophical Grammar*. On the basis of the definition of addition by recursion given above, Skolem's proof by induction of the law of associativity is in two steps:

(1) If $c = 0$, then we have $(a + b) = (b + a)$, since $a + 0 = a$
(2) If we suppose that associativity holds for a value c, then we have:
$(a + b) + (c + 1) = ((a + b) + c) + 1$
$(a + b) + (c + 1) = (a + (b + c)) + 1$
$(a + b) + (c + 1) = a + ((b + c) + 1)$
$(a + b) + (c + 1) = a + (b + (c + 1))$

This shows that, for any a and b, associativity holds for $c + 1$ (Skolem 1967b: 305–6). Of this proof, Wittgenstein pointed out that: ' "$a + (b + c) = (a + b) + c$" . . . $A(c)$ can be construed as a basic rule of a system. As such, it can only be laid down, but not *asserted*, or denied (hence no law of excluded middle)' (*PR*, § 163). He was reiterating here his account of proofs by induction: they do not assert anything. He also pointed out that a 'statement' of the law of asso-

attempt to eliminate the quantifiers and express the same sort of generality that Russell and Whitehead had endeavoured to capture in quantified propositions with "functional assertions" containing unbounded variables. Such an exercise, Wittgenstein believed, would merely replace one set of problems with another' (Shanker 1987: 200). (I do not see any textual evidence supporting this last claim, nor do I see, once more, an 'offensive' again Skolem.) It seems to me that it is Shanker who has confused generality in Skolem's primitive recursive arithmetic and generality in the theory of classes. Shanker further misreads both authors when he claims that the following remark by Wittgenstein is a criticism of Skolem: 'If we want to see what has been proved, we ought to look at nothing but the proof. We ought not to confuse the infinite possibility of its application with what is actually proved. The infinite possibility of application is *not* proved! The most striking thing about a recursive proof is that what it alleges to prove is not what comes out of it' (*PR*, § 163). Wittgenstein is trying here to block an extensionalist account of the proof, which is described in this passage as 'the usual mistake in confusing the extension of its application with what it genuinely contains' (*PR*, § 163), and this is precisely what Skolem also wanted to avoid. Otherwise, his paper would indeed have been only a futile 'exercise'. I thus see absolutely no basis for this further claim by Shanker: 'What Skolem's argument shares with Cantor's . . . is the transgression of the logico-syntactical boundary between finite totalities and infinite series, from which follows such confused assumptions as that what is shown by induction can be counted, quantified over, or asserted in a "general proposition" ' (Shanker 1987: 205). (Shanker refers here to Cantor's use of the diagonal method in the proof of the undenumerability of the real numbers; see section 7.1 for details.)

ciativity does not even appear in the proof. So how can one say that this is a proof of the law? A simple solution consists of replacing the variables by specific numbers and calculating. For Wittgenstein, it is the possibility of working out such examples which makes the inductive proof a proof of the law of associativity:

> I want to say: It is only in the sense in which you can call working out such an example a proof of the algebraic proposition that the proof by induction is a proof of the proposition. Only to that extent is it a check of the algebraic proposition. (It is a check of its structure not of its generality). (*PG*, p. 396)[37]

Perhaps the most interesting of Wittgenstein's many comments on this proof consisted in rewriting it, replacing induction by the rule of inference:

$$F(0) = G(0),$$
$$F(Sx) = B(F(x)),\ G(Sx) = B(G(x)) \vdash F(x) = G(x)$$

(*PG*, pp. 397, 414). This is a rule affirming the uniqueness of a function defined by recursion, a notion which was deemed by Louis Goodstein, who had been his student in the early 1930s, to be 'far more intuitively acceptable' (Goodstein 1972: 281). Goodstein was actually deeply influenced by Wittgenstein. In the preface to his book *Constructive Formalism*, Goodstein acknowleged his debt to Wittgenstein:

> Of the many friends who have helped, encouraged and inspired this work, first and foremost I must mention Ludwig Wittgenstein, to whose lectures in Cambridge between 1931–34 and the many conversations I was privileged to have with him, I am immensely indebted; only in recent years have I grown to understand how much he taught me. (1951: 10)

Goodstein also rejected classical quantification theory, and he developed in the 1940s another well-known system of quantifier-free arithmetic, the equational calculus. In his first paper on the subject, 'Function Theory in an Axiom-Free Equational Calculus', written in 1941 but published in 1945, Goodstein went even further than Skolem by dispensing altogether with the propositional calculus and mathematical induction and developing a purely equational calculus in which all propositions are equations of the form F = G, where F and G are primitive recursive functions or terms and where the rules are

[37] See also *PR* § 163.

substitution and uniqueness instead of induction.[38] This calculus is 'logic-free': sentential connectives and bounded quantifiers are introduced arithmetically. It is based on the notion of uniqueness (as opposed to mathematical induction), an idea he attributed to Wittgenstein (Goodstein 1972: 281).[39] Goodstein even alluded in print to Wittgenstein's remarks as early as 1945, in a footnote to his paper on 'Function Theory in an Axiom-Free Equational Calculus' (1945: 407 n.). This is one interesting example of a positive development emerging from Wittgenstein's remarks on mathematics. I do not wish to imply, however, that Goodstein's equational calculus was something like the necessary logical continuation of Wittgenstein's ideas. The equational calculus was simultaneously discovered independently of Wittgenstein by Haskell Curry (1941). (It is worth noticing, however, that Curry was also building on work by Skolem and by Hilbert and Bernays.) Rather one should emphasize that, inasmuch as replacing induction by uniqueness, etc., is an improvement on Skolem's primitive recursive arithmetic, this is an indication of the depth of Wittgenstein's thoughts on the topic. Wittgenstein's remarks on induction are not, however, a critique of the method itself but rather of the language in which it is presented (Goodstein 1972: 281).

It is also interesting to point out that in Goodstein's equational calculus there are means to express bounded universal and existential propositions—which are equivalent to a finite number of conjunctions and disjunctions—but, as in the case of Skolem's primitive recursive arithmetic, negations of unbounded universal propositions cannot be expressed (Goodstein 1951: 30–1). As far as generality is concerned, Goodstein's equational calculus also quite well embodies Wittgenstein's ideas. I need only cite here the last sentence of Goodstein's 1945 paper: 'Generality in the calculus may be exhibited without the use of variable signs, the generality of the theorem showing itself in the generality of the proof, but a calculus which contains no variable signs can only *show* generality, not *express* it' (1945: 434). (By 'variable' Goodstein means here what Russell would call 'apparent variables'.)

[38] See Goodstein (1971: ch. 7) for an introduction to his equational calculus. I should add that the essential idea that every recursive formula can be brought into an equational form F = 0 (with F containing only symbols for primitive recursive functions) is due to Gödel in his famous 1931 paper on incompleteness (Gödel 1967).

[39] This is actually an idea which originates in Dedekind's *Was sind und was sollen die Zahlen* (1963b: § 132) and which was 'rediscovered' by category theorists in the 1960s. See Lawvere (1964).

It should be clear that writings of the 1929–33 period indicate that Wittgenstein had abandoned his earlier view of the universal quantifier as a logical product and the existential quantifier as a logical sum, and that this led him to an even more radical critique of classical quantification theory than the intuitionist one. If Goodstein's work is a good indication of the direction taken by Wittgenstein's ideas in logic in the 1930s, as I believe strongly that it is, their standpoint is indeed even more radical than that of intuitionism, since they dispense with quantification theory altogether. Intuitionists simply refrain from asserting the Law of Excluded Middle in certain contexts. So, although intuitionists would refuse to apply the Law of Excluded Middle to negations of propositions involving quantification over the natural numbers, intuitionism still contains such negations, while they simply cannot be expressed in Skolem's primitive recursive arithmetic or in Goodstein's equational calculus. It is syntactically impossible, since universality is expressed only by the use of a free variable. Intuitionism is therefore wider than these and further in spirit from Wittgenstein's finitism.

5

From Truth-Functional Logic to a Logic of Equations

5.1. LOGICAL FORM AND COLOUR EXCLUSION

In order to appreciate fully the depth of Wittgenstein's commitment to finitism, one ought to look for the causes of his change of mind on quantification. They are to be found in underlying changes in the notion of analysis which took place in the first steps away from the *Tractatus Logico-Philosophicus* in the manuscripts of 1929.[1] Upon his return to Cambridge in January of that year, Wittgenstein immediately started writing extensively. This was a period of intense intellectual fervour. Early entries in his notebooks indicate that he had started to think afresh some of the problems even before his arrival in Cambridge and that he moved in new directions very quickly. For example, on page 17 of MS 105 (towards the end of February) he wrote: 'One could surely replace the logic of tautologies by a logic of equations' (*WA* i, p. 7). This remark indicates that Wittgenstein had already given up or was about to give up one of the central claims of the *Tractatus Logico-Philosophicus*, that of the completeness of its truth-functional logic, which is expressed in proposition 6. Wittgenstein had had discussions in 1928 with members of the Vienna Circle, but no records of these were kept and no manuscripts from that year—if there were any—survive. Therefore, in order to understand the steps leading to a complete dismantlement of the *Tractatus Logico-Philosophicus*, we must concentrate on the problems that Wittgenstein raised in his manuscripts, early in 1929. One such problem is the exclusion of colours. It was in order to present his new solution to it that he wrote, again early in 1929, the paper 'Some Remarks on the Logical Form'. He had put forward in the

[1] This is also a claim made by Michael Wrigley (1995: 166): 'Wittgenstein's interest in the infinite first arose as a result of questions which had nothing to do with the philosophy of mathematics'.

Tractatus Logico-Philosophicus a symbolism, the T-F notation, which was meant to show that all propositions are truth-functions of contingently true or false elementary propositions and also to exclude nonsensical constructions. But the syntax of ordinary language is, Wittgenstein tells us, 'not quite adequate to this purpose' since 'it does not in all cases prevent the construction of nonsensical pseudo-propositions' (SRLF, p. 162).[2] For example, the syntax of ordinary language allows the formulation of propositions such as '*x* is red and *x* is green', with *x* designating the same patch, which is plainly false since—the claims of some modern psychologists and naturalist philosophers notwithstanding—nothing can be both red and green at the same time. For Wittgenstein it was the task of an adequate logical notation to block the construction of such nonsensical structures, but he knew all along that his T-F notation would allow the construction of propositions of the form *p* & *q*, such as '*x* is red and *x* is green'. Indeed, if such a proposition is false, then it is necessarily false and, since all necessity is in the *Tractatus Logico-Philosophicus* logical necessity, its necessary falsehood would have implied that it is a logical contradiction. But, as is pointed out at the end of 'Some Remarks on Logical Form', '*x* is red and *x* is green' cannot be shown to be a logical contradiction in the T-F notation. Indeed, an attempt at representing it as a contradiction would look like this:

x is red	*x* is green	*x* is red & *x* is green
T	T	F
T	F	F
F	T	F
F	F	F

This attempt is unsuccessful, however, since the first line, TTF, represents an impossible assignment of truth-possibilities or, as Wittgenstein said it, 'it is nonsense' because it 'gives the proposition a greater logical multiplicity than that of the actual possibilities' (SRLF, p. 170). A solution to this problem was briefly alluded to in proposition 6.3751. This section is worth quoting in full:

> For two colours, e.g. to be at one place in the visual field, is impossible, logically impossible, for it is excluded by the logical structure of colour.

[2] By syntax, Wittgenstein meant 'the rules which tell us in which connections only a word gives sense, thus excluding all nonsensical structures' (SRLF, p. 162).

Let us consider how this contradiction presents itself in physics. Somewhat as follows: That a particle cannot at the same time have two velocities, i.e. that at the same time it cannot be in two places, i.e. that particles in different places at the same time cannot be identical.

(It is clear that the logical product of two elementary propositions can neither be a tautology nor a contradiction. The assertion that a point in the visual field has two different colours at the same time, is a contradiction.) (*TLP*, 6.3751)

It seems fairly obvious from this paragraph that Wittgenstein was aware of the fact that *p* & *q* cannot be turned into a contradiction in the T-F notation. It is legitimate to assume, however, that he thought that his solution would nevertheless show that '*x* is red and *x* is green' is a contradiction. But what was the original solution? It seems that Wittgenstein thought he could turn colour exclusion into a logical contradiction by taking propositions ascribing degrees to properties as complexes to be analysed further into a logical product of elementary propositions. Then, in a manner analogous to the case of particles in kinematics (where they cannot have two velocities at the same time) which he did not specify, Wittgenstein thought he could show that a given patch cannot have two colours at the same time.

The true nature of Wittgenstein's solution is a matter of speculation. But, if Wittgenstein wrote 'Some Remarks on Logical Form' in 1929 to present a new solution, it means that he was no longer in agreement with this alleged solution. What was its blemish? Frank Ramsey had already expressed his dissatisfaction with Wittgenstein's alleged solution in his critical notice of the *Tractatus Logico-Philosophicus*, written in 1923 (Ramsey 1923: 473). One also finds Ramsey referring to proposition 6.3751 in a posthumous fragment:

We know that '*x* is both red and yellow' is nonsense. It does not express a thought. To recognize nonsense, etc. is a gift of nature.

The analysis of thought *must* make this clear and this must guide us in analysing thought. Wittgenstein does not really explain this. (Ramsey 1990: 59)

We have already seen that Wittgenstein and Ramsey had frequent conversations in 1929. It is possible that Wittgenstein was put under pressure on this point by Ramsey during their conversations.

We find Wittgenstein realizing that 'statements of degree', as he would call them in 'Some Remarks on Logical Form' (p. 168), are unanalysable, for reasons which should become clear in a moment.

Since statements of degree are unanalysable, numbers have to enter in elementary propositions, and once there are numbers, propositions cannot be independent. Wittgenstein was thus led to the abandonment of one of the pillars of the *Tractatus Logico-Philosophicus*, the thesis of the logical independence of elementary propositions (*TLP*, 5.134; SRLF, p. 168). With this thesis gone, it was inevitable that the whole logic of the *Tractatus Logico-Philosophicus* would fall to the ground. Accordingly, and rightly so, commentators have given a prominent place to the problem of colour exclusion in their account of the dismantling of the *Tractatus Logico-Philosophicus* during the years 1929–33.[3] The surroundings of this crucial problem are therefore worth as thorough an investigation as possible. As I see it, Wittgenstein's changing thoughts on logical form (of elementary propositions) constitute the proper context in which to place the problem of colour exclusion. It is also from this standpoint (that of his investigations on elementary propositions) that we can clearly see Wittgenstein conceiving first of a 'phenomenology'—a term borrowed from Mach and Boltzmann[4]—as an attempt to build anew the *Tractatus Logico-Philosophicus*, and finally realizing that his attempt was resting on absurd assumptions (about elementary propositions or, as he would call them by then, 'phenomenological assertions') that had to be given up. At that point, he definitively moved away from logical atomism towards the analysis of ordinary language which characterizes his later philosophy.[5]

The idea of 'analysis', from complex to simple, contained in logical atomism forced Wittgenstein to postulate in the *Tractatus*

[3] See e.g. Kenny (1973) or Hacker (1986: ch. 5).

[4] There are similarities between Wittgenstein's and Husserl's phenomenology, over and above their use of the word 'phenomenology' to describe their work. The relations between the two have been discussed extensively in the literature, from Spiegelberg's early paper (1968) to Hintikka's recent work (1990). I am not interested in such parallels here.

[5] I am largely in agreement with the reconstruction of the evolution of Wittgenstein's thought in 1929 proposed by Stern (1991). The following could serve as a summary here: 'At first, [Wittgenstein] retained the Tractarian conviction that language is grounded on reference to objects, which he now identified with the contents of experience. This project of analysing the structure of the experientially given is briefly articulated in the paper "Some Remarks on Logical Form". At this point, in the early months of 1929, he conceived of the project as a matter of articulating a "phenomenological language", a language for the description of immediate experience. Later that year, he gave up the idea that philosophy ought to start from a description of the immediately given, motivated by the conviction that philosophy must begin with the language that we ordinarily speak' (p. 205).

Logico-Philosophicus the existence of a level of elementary (or atomic) propositions: analysis must come to an end, otherwise the truth of any given proposition would rely on the truth of another proposition. This bottom level will be that of elementary propositions, which must therefore be logically independent from each other (*TLP*, 5.134). These elementary propositions must also consist of a combination 'simple names' which refer to 'simple objects' (*TLP*, 4.221), i.e. 'the ultimate connection of the terms, the immediate connection which cannot be broken without destroying the propositional form as such' (SRLF, pp. 162–3). That there must be such a level is simply a consequence of the Russellian idea of 'analysis', and one does not need to know the precise nature of simple objects. They are required in order to get the 'analysis' going (*NB*, p. 60) and the major theses of the *Tractatus Logico-Philosophicus*, such as the picture theory, are stated independently of any categorical specification of the simple objects. Wittgenstein's notorious silence on this question has led to an exegetical debate. Some crude readings of the *Tractatus Logico-Philosophicus* can at least be set aside: simple objects cannot be simply 'material points' in the sense of Hertz[6] nor simply 'sense-data' in the sense of Russell (or Moore).[7] What is not open to controversy is the fact that Wittgenstein assumed in the *Tractatus Logico-Philosophicus* that there is a level of elementary propositions and that it is the most fundamental level of our language. Since we can assume, without further inferences, nothing about the nature of simple objects, apart from the fact that they are named by simple names, we can say very little at the outset about the structure of elementary propositions. Unsurprisingly, we find Wittgenstein claiming that we cannot foresee their form: since we cannot say anything about simple objects, we cannot say anything about their names and, since 'we cannot give the number of names with different meanings, we cannot give the composition of the elementary proposition' (*TLP*, 5.55). In fact, Wittgenstein claimed, 'we have a concept of the elementary proposition apart from its special logical form' (5.555). Wittgenstein also said that 'logical forms are *anumerical*' and that there are in logic 'no pre-eminent numbers' (*TLP*, 4.128). It is from

[6] Cf. *NB*, p. 67 and Hertz (1899: §§ 9–11). On the influence of Hertz, see Anscombe (1971: pp. 25–8) or Griffin (1964: pp. 5, 99–102, 150).

[7] To my mind, a proper understanding of the Russellian sources of the *Tractatus* shows that the simple objects are akin to, but different in fundamental respects from, Russell's 'objects of acquaintance'. Such an interpretation was set forth in Hintikka and Hintikka (1986: ch. 3).

this standpoint that he criticized Russell's conceptions in *Principia Mathematica* (*TLP*, 5.553–5.5542) and, later, Carnap's in *The Logical Structure of the World* (*WVC*, p. 182; *PG*, p. 210).

Although the logical form of elementary preposition is unforeseeable, Wittgenstein had no choice but to assume they had some 'form'; otherwise, he would not have been able to even write them down. So we find him in 4.24 presenting elementary propositions in a rather Russellian manner:

> The names are the simple symbols, I indicate them by single letters (x, y, z).
> The elementary proposition I write as a function of the names, in the form 'fx', '$\phi(x, y)$', etc. Or I indicate it by the letters p, q, r. (*TLP*, 4.24)

This notation makes much sense, since elementary propositions are said to be pictures of states of affairs (*Sachverhalten*), where simple objects are combined together: simple objects being named by simple names, the elementary propositions will take the form of *functions* of simple names, thus picturing the combination of simple objects into given states of affairs. With this minimal assumption about the 'form' of elementary propositions, Wittgenstein was able to get his truth-functional apparatus going, without having to know in advance whether elementary propositions consist of dyadic or 27-termed relations!

In 1929, Wittgenstein was ready to throw even these minimal assumptions overboard. Although his remarks on this subject remain obscure, it seems that he envisaged in the *Tractatus Logico-Philosophicus* a two-stage analysis: first, there are specific spaces, such as colour space or sound space, within which some constructions could be shown to be impossible. Secondly, there is an all-encompassing logical space. In it, all other forms (those specific to colour, space, etc.) are, so to speak, subsumed under the logical form. In a passage originating in MS 106, Wittgenstein speaks analogously of 'projections'. He appears to be struck by the vacuity of the logical form (here the subject–predicate form, but this remark applies equally to Frege's function-theoretic forms), which he sees as uninformative as a projection which would transform a variety of shapes on a given plane into one unique shape, namely circles (of varying size), on a second plane:

> Imagine two planes, with figure on plane I that we wish to map on plane II by some method of projection. It is open to us to fix a method of projection (such as orthogonal projection) and then to interpret the images on plane II

according to this method of mapping. But we could also adopt a quite different procedure: we might for some reason lay down that the images on plan II should all be circles, no matter what the figures on plane I may be. That is, different figures on I are mapped on to II by different methods of projection. In order in this case to construe the circles in II as images, I shall have to say for each circle what method of projection belongs to it. But the mere fact that a figure is represented on II as a circle will say nothing.—It is like this with reality if we map it onto subject–predicate propositions. The fact that we use subject–predicate propositions is only a matter of our notation; the subject–predicate form does not in itself amount to a logical form and is the way of expressing countless fundamentally different logical forms, like the circles on plane II. The forms of the propositions: 'The clock is round', 'The man is tall', 'The patch is red', 'The picture is pretty', have nothing in common. (*WA* i, p. 63)

This crucial passage was to be repeated not only in the paper 'Some Remarks on Logical Form' (pp. 164–5) but also in *Philosophical Remarks* (§ 93) and *Philosophical Grammar* (pp. 204–5). This is a strong indication of its importance for Wittgenstein. He seems to have realized that the subject–predicate form is, so to speak, a generic 'disguise' under which we put different kinds of propositions. The differences between these propositions are then masked by their common subject–predicate form, and the least interesting aspect of these many propositions is precisely that they have such a common form; the fact that they share a common form 'will say nothing'. This is the first occurrence of a fundamental idea of the later Wittgenstein. It is captured by a quotation from Shakespeare's *King Lear*, which he intended to put as epigraph for the *Philosophical Investigations*: 'I'll teach you differences.'[8] Now, since Wittgenstein thought that Frege's distinction between object and concept was a generalization of the subject–predicate form, from this he concluded that it was not a single logical form either (*WA* i, p. 64; *PR*, § 93).[9]

[8] For the anecdote, see Rhees (1984: 157). The quotation occurs in Act I, sc. iv, when Kent scolds Oswald for not showing respect to King Lear. The 'differences' are here differences of rank and presumably not, however, the kind of essential differences emphasized by Wittgenstein.

[9] It might be argued that Wittgenstein's claim that Frege's notions of concept and object are simply generalizations of the subject–predicate form is mistaken. To begin with, Frege thought that the subject–predicate form was linked to ordinary language and carried with it deficiencies which would be avoided in the logical notation by adopting the distinction between concept and object. I cannot discuss this issue here; my aim is merely to get clear about Wittgenstein's ideas, not to evaluate them.

It dawned on Wittgenstein that neither the subject–predicate form nor the logical form of the *Tractatus Logico-Philosophicus* was fundamental: under them are put all sorts of propositions whose form are in fact different. Therefore, in order to achieve a correct analysis of elementary propositions, one must go through the logical form to get to the various underlying forms. This is a crucial step away from the doctrines of the *Tractatus Logico-Philosophicus*. Recall that the point of the book was to show the essential pictorial nature of all propositions. This was done in two stages. Starting from the thesis that elementary propositions are pictures of possible (or actual) states of affairs, Wittgenstein was able to extend his account to complex propositions by introducing truth-functions. With the introduction of the N operator, he was able to obtain all other truth-functional connectives, and the task of explaining how the pictoriality of elementary propositions could be extended to complex propositions was thus reduced to that of showing that a conjunction of pictures remains a picture and that the negation of a picture also remains a picture. Wittgenstein was thus in a position to claim that pictoriality is the essence of language.

This idea is behind the completeness claim embodied in proposition 6. It is this very completeness of truth-functional logic, to which Wittgenstein referred in the passage quoted above as the 'logic of tautologies', which is abandoned in 1929. This is not, however, in itself a sufficient reason for believing that he had rejected the picture theory. As I read him, he was simply saying that the whole logical apparatus of the *Tractatus Logico-Philosophicus* is only capturing *part* of the set of propositions. The picture theory still stands for these propositions; it has only lost its generality (and with this the very idea of a search for the unique essence of language is jeopardized). Wittgenstein never denied that one can give the general form of a proposition, but simply claimed that it 'determines a proposition as part of a calculus' (*PG*, p. 125). According to him the general form of a proposition given in proposition 6 of the *Tractatus Logico-Philosophicus* was simply a 'handle for the truth-functions', an expression 'which shows us what part of grammar comes into play' (p. 124). This is further confirmed by a passage from *Philosophical Investigations*: 'Asked what a proposition is . . . we shall give examples and these will include what one may call inductively defined series of propositions. *This* is the kind of way in which we have such a concept as "proposition". (Compare the concept of a proposition with the concept of number.)' (*PI*, § 135) Here, Wittgenstein claims that the concept of proposition includes but is not equal to 'inductively defined series of propositions'. The reference to proposition 6 of the *Tractatus Logico-Philosophicus* is glaringly obvious.[10]

The abandonment of the completeness claim means that another 'logic' is underlying our ordinary language. It seems that this 'logic' is the 'logic of equations' mentioned in the very first quotation of this section, and that this 'logic' is not Russellian but precisely the new logic of 1929 presented in section 4.2 above. It is worth remembering here that that which is not captured by this logical scheme is neither of a pictorial nature nor a fully-fledged proposition but rather an 'appearance' of proposition or pseudo-proposition (*Scheinsatz*), and that equations of mathematics are precisely pseudo-propositions (*TLP*, 6.2). The truth-functional logic of the *Tractatus Logico-Philosophicus* does not cover mathematical pseudo-propositions. These are 'logic-free'. Wittgenstein often spoke as if the 'logic of tautologies' were embedded in this new 'logic' or 'grammar'. For example, in discussion with Schlick and Waismann, Wittgenstein stated clearly that his truth-functional logic was only part of 'a more comprehensive syntax': 'What I at first paid no attention to was that the syntax of logical constants forms only part of a more comprehensive syntax. Thus I can, for example, construct the logical product $p.q$ only if p and q do not determine the same co-ordinate twice' (*WVC*, p. 77). The same claim is made in *Philosophical Remarks*: 'The rules for "and", "or", "not" etc., which I represented by means of the T-F notation, are *a part* of the grammar of these words, but not *the whole*' (§ 83). But not only was truth-functional logic by then seen as only part of a more comprehensive new 'logic', it was also found by Wittgenstein to be wholly inadequate at the level of elementary propositions. We have already seen that this much is precisely shown by the new treatment of the problem of colour exclusion.

In Russell's logical atomism, the notion of object is related to the subject–predicate or the function–argument form. But, as Wittgenstein pointed out to Schlick and Waismann during a meeting at the end of 1929, an entirely different description of our immediate experience can be achieved by analytic means—Hertz's *Principles of Mechanics* comes to mind—without ever mentioning 'objects' or their (spatial)

[10] Therefore, in order to sustain the claim that Wittgenstein effectively abandoned the picture theory, one needs further arguments. This question cannot, however, be dealt with here.

relations. Here, there is no function–argument form, no 'logic of relations':

> When Frege and Russell spoke of objects they always had in mind things that are, in language, represented by nouns, that is, say, bodies like chairs and tables. The whole conception of objects is hence very closely connected with the subject-predicate form of propositions. It is clear that where there is no subject-predicate form it is also impossible to speak of objects in this sense. Now I can describe this room in an entirely different way, e.g. by describing the surface of the room analytically by means of an equation and stating the distribution of colour on this surface. In the case of this form of description, single 'objects', chairs, books, tables, and their spatial positions are not mentioned any more. Here we have no relation, all that does not exist. (*WVC*, pp. 41–2)

One finds another striking condemnation in Waismann's 'Theses':

> The proposition 'Orange lies between yellow and red', for example, sounds just like the proposition 'This table stands between this chair and that window' and that is why it so easily appears as though such a proposition described those colours. Here the use of the noun form continously misleads us.
>
> The same holds of propositional functions. The symbol '*fx*' is taken from the case in which '*f*' signifies a predicate and '*x*' a variable noun. When we advance to elementary propositions propositional functions (classes) become worthless. (*WVC*, p. 258)

It is clear that many aspects of the *Tractatus Logico-Philosophicus* were thought of while Wittgenstein had in the back of his mind the sort of analysis provided by physics, more precisely the sort of analysis found in Heinrich Hertz's *Principles of Mechanics*. The above passages are some of the clearest examples of the underlying influence of Hertz. Indeed, if we assume that Wittgenstein had in mind the type of coordinate analysis which he would have found in the *Principles of Mechanics*, his remarks make perfect sense. If the logical form of the *Tractatus Logico-Philosophicus* is worthless at the level of elementary propositions, these must have a different form, and if the above remarks are of any value, this form will not be unlike that of the equations of physics (mechanics). Quite unsurprisingly, we find Wittgenstein entertaining in the very first pages of MS 105 the idea that physics is the 'true phenomenology' (*WA* i, p. 4) and telling Schlick and Waismann:

> The logical structure of elementary propositions need not have the slightest similarity with the logical structure of propositions.

Just think of the equations of physics—how tremendously complex their structure is. Elementary propositions, too, will have this degree of complexity. (*WVC*, p. 42)

Contrary to what Russell believed and stated in his introduction, Wittgenstein was not concerned in the *Tractatus Logico-Philosophicus* 'with the conditions which would have to be fulfilled by a logically perfect language' (*TLP*, p. 7). According to him, ordinary language was in perfect logical order (*TLP*, 5.5563). He was rather trying to unravel the underlying logic that every language, including ordinary language, must possess in order for it to be able to provide pictures of the world. So, while Russell, being concerned with a logically perfect language and having essentially in mind its application to mathematics, could have dismissed the above objections as irrelevant, Wittgenstein saw the attempt at subsuming many diverse forms under one general 'logical form' as seriously flawed. The point I wish to emphasize is that the problem of colour-exclusion should be seen as an argument backing up this fundamental change of perpective.

As mentioned earlier, the key to the paper 'Some Remarks on Logical Form' is the realization that so-called 'statements of degree' are unanalysable. What does it mean? In the *Tractatus Logico-Philosophicus* 'analysis' was conceived as decomposition from complex to simple; and, since complex propositions are concatenations of elementary propositions by means of truth-functional operators, to say that 'statements of degree' are analysable means therefore that they are logical products of even more elementary propositions. (Wittgenstein also thought that such products should be accompanied, when not dealing with a class 'defined by grammar', by a 'concluding proposition', which says something like 'and these are the only so and so'.)[11] In a passage appearing for the first time in MS 106

[11] For example, supposing that there are three men in a room, Schlick, Waismann, and Wittgenstein, respectively a, b, and c, and that they are all wearing trousers. The formalized version of 'All men in this room are wearing trousers' would not just read something like: $M(a) \rightarrow T(a)$ & $M(b) \rightarrow T(b)$ & $M(c) \rightarrow T(c)$. It would also need a 'concluding proposition' to be tagged at the end saying that for all x it could not be the case that x is a man and that x is neither a nor b nor c. The need for such a 'concluding proposition' has originally to do with Russell's distinction between particular and general facts (e.g. in *The Philosophy of Logical Atomism* (Russell 1985: 101 f.)). Although for Wittgenstein there are no general facts and such 'concluding propositions' could not be written in the notation of the *TLP*—Ramsey had defended his account of general propositions on this point (1978: 56–7)—Wittgenstein seemed to maintain the need, in cases where the domain is not defined by grammar, for such

(*WA* i, p. 55) and reproduced in *Philosophical Remarks*, Wittgenstein realized that it makes no sense to analyse statements about degrees in terms of logical product of elementary propositions:

And different degrees of red are incompatible with one another. Someone might perhaps imagine this being explained by supposing that certain small quantities of red added together would yield a specific degree of red. But in that case what does it mean if we say, for example, that five of these quantities of red are present? It cannot, of course, be a logical product of quantity no. 1 being present, and quantity no. 2, etc., up to 5; for how would these be distinguished from another? Thus the proposition that 5 degrees of red are present can't be analysed like this. Neither can I have a concluding proposition that this is all the red that is present in this colour: for there is no sense in saying that no more red is needed, since I can't add quantities of red with the 'and' of logic. (*PR*, § 76)

Wittgenstein then offers another example, that of a rod which is 3 metres long:

Neither does it mean anything to say that a rod which is 3 meters long is 2 meters long, because it is 2+1 meters long, since we can't say it is 2 meters long and that it is 1 meter long. The length of 3 meters is something new. (*WA* i, p. 55; *PR*, § 76)

It is indeed difficult to see how a proposition such as 'This rod is 3 metres long' is analysable further into the conjunction:

'This rod is 2 metres long' & 'This rod is 1 metre long'

This would mean that facts corresponding to the logically independent elementary propositions 'This rod is 2 metres long' and 'This rod is 1 metre long' also obtain independently,[12] and this makes no sense whatsoever: if the rod is indeed 3 metres long, it is not at the same time also 2 metres long and 1 metre long. This extremely important albeit obvious thought led Wittgenstein to a further even more important one: it makes no more sense to say that this is all that it is, since the fact that it is 3 metres long also implies that it is not 4 metres long, and so forth . . . (A similar argument is at the core of the paper 'Some Remarks on Logical Form' (pp. 167–8).) In this very passage, Wittgenstein realizes for the first time that, for the same

'concluding propositions'. See Wittgenstein's own explanations in *WVC*, pp. 44–5, 51–3, and *M*, pp. 297–99.

[12] Recall that in the *Tractatus* one cannot infer from the existence of one state of affairs the existence of another state of affairs (*TLP*, 2.062). See also *TLP*, 5.134, 5.135.

reasons, there must be 'constructions' in language outside the scope of truth-functional logic:

> And yet I can say, when I see two different red-blues: there's an even redder blue than the redder of these two. That is to say, from the given I can construct what is not given. . . .
>
> That makes it look as if a construction might be possible within the elementary proposition. That is to say, as if there were a construction in logic which didn't work by means of truth functions.
>
> What's more, it also seems that these constructions have an effect on one proposition's following logically from another.
>
> For if different degrees exclude one another it follows from the presence of one that the other is not present. In that case, two elementary propositions can contradict each other. (*WA* i, pp. 55–6; *PR*, § 76)

We can see here Wittgenstein realizing that there are other operations in language, not just the truth-functions. The idea of the 'completeness' of the truth-functional logic is thus shattered. Wittgenstein is now forced to recognize that what he said in the *Tractatus Logico-Philosophicus* 'doesn't exhaust the grammatical rules for "and", "not", "or" etc.' (*PR*, § 82), and to abandon the logical independence of elementary propositions, the cornerstone of the logic of the *Tractatus Logico-Philosophicus*.

We are now in a position to appreciate fully the central role that the notion of 'hypothesis' (to be introduced below) was to play during the transitional period: 'hypotheses' were thought by Wittgenstein to be precisely those operations by means of which constructions could take place in language outside the scope of truth-functional logic. The 'logic' which seems to be at work here appears to be precisely the 'logic of equations' referred to in the first quotation of this section, since this expression obviously refers to arithmetic, which did not admit of a truth-functional treatment in the *Tractatus Logico-Philosophicus*. Moreover, numbers (rational or irrational) now enter in elementary propositions (SRLF, p. 165; *WVC*, p. 42). This is a new development: in the *Tractatus Logico-Philosophicus* numbers were described as 'exponents of an operation' (*TLP*, 6.021) and as such could not have appeared in any elementary propositions. That numbers must enter in elementary propositions is, however, a fairly obvious consequence of the fact that so-called 'statements of degree' are not analysable further: since such statements cannot be broken into further more elementary propositions, the multiplicity of the phenomena will not be captured by the use of the conjunction, as the above truth-table showed, and the proposition attributing a given degree will have to be elementary. In particular, it will have to contain in itself the same multiplicity as that of the given degree, and the use of numbers (e.g. to give the values of coordinates) seems necessary for the expression of that multiplicity (SRLF, p. 168).

The above reasoning showing that 'statements of degree' are unanalysable seems so obvious that one might wonder, especially in light of the analogy with kinematics in proposition 6.3751, why Wittgenstein did not see at the time of writing the *Tractatus Logico-Philosophicus* that numbers would have to enter in elementary propositions or, as he would put it in Hertzian terms, that the value of the coordinates had to be determined. Wittgenstein himself admits:

> The concept of an 'elementary proposition' now loses all of its earlier significance. . . .
>
> In my old conception of an elementary proposition there was no determination of the value of a co-ordinate; although my remark that a coloured body is in a colour-space, etc., should have put me straight on to this. (*PR*, § 83)

This vexed question need not detain us here. For our purposes it suffices that we recognize that Wittgenstein ultimately believed that it was necessary to introduce numbers in elementary propositions and that he had, as a result, to abandon the thesis of their logical independence. This realization has some important consequences. For one thing, we can now see with much clarity the source of Wittgenstein's finitist pronouncements in philosophy of mathematics from 1929 onwards. He had to find new operations for the contructions in language which were not captured by the N operator. He had already excluded arithmetic from the scope of logic in the *Tractatus Logico-Philosophicus* (which was decidedly *not* a contribution to logicist foundations), and, as he candidly admits, he found himself 'thrown back to arithmetic' against his will (*WA* i, p. 7). If arithmetical operations explain constructions in language outside the scope of truth-functional logic, they ought not to be reduced in turn to logical operations, as logicists would have it, because the very point of having avoided truth-functional logic in the first place would be lost. Accordingly, Wittgenstein repeatedly expressed the view that general arithmetical propositions (with the quantifier ranging over the natural numbers) can only be 'correctly expressed by means of induction', not by a statement, which would be of the form $\forall x\, F(x)$. The point is of importance: one sees here another argument for Wittgenstein's rejection of logicism, which is linked—since it implies that there is no room for infinite disjunctions or conjunctions—to his finitist account of arithmetic.[13] As we saw in section 4.2 above, Wittgenstein changed his mind in 1929 about quantification and he was thus led to positions very close to Skolem's finitism, from which Goodstein derived his 'logic-free' equational calculus. When Wittgenstein spoke of a 'logic of equations' (*WA* i, p. 7), he clearly had some such calculus in mind. The introduction of numbers in elementary propositions did also play a crucial role in the downfall of the 'phenomenology'. Indeed, as a result of this new situation Wittgenstein's notion of 'analysis' was to undergo rapid and fundamental transformations, and the train of thoughts set in motion by the reasoning presented above was to lead ultimately to the abandonment of the very project of logical atomism, although not before Wittgenstein tried to think the whole matter afresh.

Wittgenstein concluded his paper 'Some Remarks on Logical Form' by saying: 'It is . . . a deficiency of our notation that it does not prevent the formation of such nonsensical constructions, and a perfect notation will have to exclude such structures by definite rules of syntax' (p. 170–1). Part of his diagnosis was that the syntax of the logical constants must be embedded in 'a more comprehensive syntax' (*WVC*, p. 77). But how are we to find out which form this 'more comprehensive syntax' will take? Wittgenstein realized that, since the logical form of elementary propositions cannot be foreseen (*WVC*, p. 42), one needs some sort of supplementary analysis of phenomena, a 'logical investigation of the phenomena themselves', in order to support the construction of a logically perfect language. In the *Tractatus Logico-Philosophicus* he had presented a conception of philosophy purified of all empirical or psychological admixture (*TLP*, 4.111, 4.112). There was no room either for epistemology, since 'the theory of knowledge is the philosophy of psychology' (*TLP*, 4.1121). Moreover, because he held such a purified conception, Wittgenstein could, without any qualms, leave questions about the nature of simple objects unanswered. To take a famous example, when Norman Malcolm asked Wittgenstein (in 1949) if he had in mind a particular type of 'simple object' when he wrote the *Tractatus Logico-Philosophicus*, the latter answered that 'at that time his thought had

[13] For a similar approach to Wittgenstein's philosophy of mathematics during the transitional period, see Hintikka (1992).

been that he was a *logician*; and that it was not his business, as a logician, to try to decide whether this thing or that was a simple thing or a complex thing, that being a purely *empirical* matter!' (Malcolm 1984: 70). The answer to such questions was considered to be a matter for the '*application* of logic' (*TLP*, 5.557) and could not have been presupposed by logic itself. At the time of writing 'Some Remarks on Logical Form', Wittgenstein still held the idea that the discovery of the form of elementary propositions is 'the task of the theory of knowledge' (pp. 162–3),[14] but the emphasis had by then shifted to the need to investigate phenomena first. Perhaps a comparison with Rudolf Carnap's 1928 *Logical Structure of the World* will help us, by way of contrast, to understand the kind of analysis now required by Wittgenstein. Indeed, Carnap also thought that his own 'constructional system', as he called it, was depending on the findings of a prior 'phenomenology':

> As concerns the content of our constructional system, let us emphasize again that it is only a tentative example. The content depends upon the material findings of the empirical sciences; for the lower levels in particular upon the findings of the phenomenology of perception, and psychology. The results of these sciences are themselves subject to debate; since a constructional system is merely the translation of such findings; its complete material correctness cannot be guaranteed. (Carnap 1967: p. 106)

Inasmuch as Carnap thought that the phenomenological inquiry was a purely empirical matter, his position did not diverge from that of the *Tractatus Logico-Philosophicus*. Wittgenstein had, however, changed his mind by early 1929, and he was by then including the theory of knowledge within the realm of 'logical' inquiry.

Wittgenstein was still holding the thesis that 'ordinary language disguises logical structure' (SRLF, p. 163), and he still believed that he could unveil it. Only this time he required, in his own words, a 'logical investigation of the phenomena themselves' which would find out their proper multiplicity, so that the logical symbolism could be adjusted to it:

> Now, we can only substitute a clear symbolism for the unprecise one by inspecting the phenomena which we want to describe, thus trying to understand their logical multiplicity. That is to say, we can only arrive at a correct analysis by, what might be called, the logical investigation of the phenomena themselves, *i.e.*, in a certain sense *a posteriori*, and not by con-

[14] See also *WVC*, p. 42.

jecturing about *a priori* possibilities . . . An atomic form cannot be foreseen. And it would be surprising if the actual phenomena had nothing more to teach us about their structure. (SRLF, pp. 163–4)

The most important feature of the notation which is meant to come out of this type of investigation is that phenomena and notation must share the same multiplicity; in other words the notation must allow all possible combinations and exclude all impossible ones. A phenomenological theory such as the *phänomenologische Farbenlehre*[15] must therefore represent perspicuously, with help of the appropriate symbolism, all the relations between colours. It must make apparent why a patch cannot be both red and green at the same time, why a reddish-yellow is possible, and so forth . . . Phenomenology is therefore concerned with *possibility* as opposed to what is *actual* and differs in that respect, according to Wittgenstein, from a physical theory. Indeed, while physics strives for *truth*, phenomenology strives for *sense* (*WA* i, p. 4); it 'describes what it has sense to say and what it has not sense to say' (*LWL*, p. 66). To Schlick and Waismann, Wittgenstein described the difference between phenomenology and physics in the following terms:

Physics wants to determine regularities; it does not set its sights on what is possible.

For this reason physics does not yield a description of the structure of phenomenological states of affairs. In phenomenology it is always a matter of possibility, i.e. of sense, not of truth and falsity. (*WVC*, p. 63)

With hindsight, we can say that this distinction between phenomenology and physics is essentially the same as that, in the *Tractatus Logico-Philosophicus*, between logic and physics. Moreover, phenomenology is, as we shall see, the precusor of the 'grammar' of the 1930s. There is a remarkable continuity here. On the other hand, phenomenology seems to differ from these other conceptions in one crucial respect. This strange form of investigation was to be at the same time 'logical' and 'a posteriori'. For this reason, there is a danger of conceiving this 'phenomenology' as a discipline lying half-way between a logical and a scientific inquiry.[16]

[15] Wittgenstein uses this expression in e.g., *PR*, § 218.

[16] See *RC*, ii, § 3. If Wittgenstein's phenomenology was such a discipline, then the empirical claim that one could see reddish-green would potentially render the whole discussion of colour exclusion otiose, and Wittgenstein's phenomenology would then appear as a piece of bad anticipatory science. The empirical claim that one could see reddish-green was made recently by psychophysiologists (Crane and Piantanida 1983)

Merrill and Jaakko Hintikka have proposed a solution to the colour-exclusion problem: define a function c which maps points in visual space into a colour space. Then the logical form of 'x is red' would be $c(x) = r$ and that of 'x is green' $c(x) = g$, and the two propositions would be logically incompatible, since the logical form of a function precludes its having two values for the same argument (Hintikka and Hintikka 1986: 123). The solution in 'Some Remarks on Logical Forms' also depended on the idea that certain functions 'can give a true proposition only for one value of their argument because—there is only room in them for one' (p. 168). According to Wittgenstein, propositions such as 'x is red' or 'x is green' are in that sense 'complete': 'That which corresponds in reality to the function . . . leaves room only for one entity—in the same sense, in fact, in which we say that there is room for one person only in a chair' (p. 169).[17] Wittgenstein spoke accordingly of the propositions 'x is red' and 'x is green' as 'colliding' in the object (SRLF, p. 169; *PR*, § 79). He became aware of a potential confusion here between 'physical' and 'logical' impossibility. Only when the later notion of grammatical rule was in place, as a result of developments to be sketched below, was Wittgenstein able to describe the confusion in clear terms. In a passage from the *Blue Book*, he referred directly to the above remark:

> 'The colours green and blue can't be in the same place simultaneously.' Here the picture of physical impossibility which suggests itself is, perhaps, not that of a barrier; rather we feel that the two colours are in each other's way. What is the origin of this idea?—We say three-people can't sit side by side on this bench; they have no room. Now the case of colours is not analogous to this; but it is somewhat analogous to saying: '3 × 18 inches won't go into 3 feet.' This is a grammatical rule and states a logical impossibility. The proposition 'three men can't sit side by side on a bench a yard long' states a physical impossibility; and this example shows clearly why the two impossibilities are confused. (*BB*, p. 56)

and Arthur Danto seized the occasion to dismiss, with much arrogance, Wittgenstein's remarks as 'anticipatory science done badly', in his introduction to C. L. Hardin's *Color for Philosophers* (Hardin 1988: xi). I would like to claim claim here, along with Alva Noë (1994: 3 n.), that it is not even clear that Wittgenstein entertained seriously the idea of a half-scientific phenomenology. If he did, he certainly abandoned it rapidly, and the entire discussion centring around the notion of grammatical rule which was to ensue has precisely nothing to do with 'anticipatory science', done badly or not. As for the empirical claim of psychophysiologists, it appears that the experiment did not pan out. So, maybe it was too soon for naturalistic-minded philosophers to claim victory (as if Wittgenstein and others were making competing claims).

[17] See also *PR*, § 77.

It has to be said, however, that even at the time of writing 'Some Remarks on Logical Form' Wittgenstein was aware of the potential confusion. His tentative solution can only be understood properly if one keeps in mind the distinction between 'sign' and 'symbol' (*TLP*, 3.32 ff.). In a nutshell, the 'symbol' is the 'sign' and its 'significant use' (*TLP*, 3.326–7). Wittgenstein expressed this idea in 'Some Remarks on Logical Form' in slightly different words: 'the sentence, together with the mode of projection which projects reality into the sentence, determines the logical form of entities' (p. 169). So, if a proposition *qua* symbol contains the logical form of the entity, then it might collide with another proposition in the same form (p. 169). The contradiction does not show in the signs but in the symbols (*PR*, § 78). So, although Wittgenstein spoke of the propositions colliding in the object and of a 'logical investigation of phenomena', he nevertheless tried to show that the contradiction between '*x* is red' and '*x* is green' is to be found in the grammar and not in the world. At approximately the time of writing his paper, he wrote in MS 106:

> Of course, this doesn't mean that inference could now not be only formal, but also material.—Sense follows from sense and so form from form. . . .
>
> 'Red and green won't both fit into the same place' doesn't mean that they are as a matter of fact never together, but that you can't even say they are together, or, consequently, that they are never together. (*WA* i, p. 58; *PR*, § 78)[18]

5.2. ASSERTIONS AND HYPOTHESES

It is now time to draw some parallels between Wittgenstein's terminology in the *Tractatus Logico-Philosophicus* and 'Some Remarks on Logical Form' and that of Heinrich Hertz in *Principles of Mechanics*. One finds in particular the idea that the logical symbolism must have the same multiplicity as the phenomena that it represents in Hertz's definition of a 'dynamical model', to which Wittgenstein referred in 4.04:

> A material system is said to be a dynamical model of a second system when the connections of the first can be expressed by such coordinates as to sat-

[18] The same remark occurs both in MS 106, therefore approximately when Wittgenstein was writing 'Some Remarks on Logical Form', and in *Philosophical Remarks*. This fact leads me to believe that Wittgenstein did not have two solutions to the problem of colour exclusion, contrary to the claim made in Austin (1980).

isfy the [condition that] the number of coordinates of the first system is equal to the number of the second. (Hertz 1899: § 418)

Wittgenstein often spoke of a 'first' and a 'second' system, expressions which were also used by Hertz in this passage; and, in conjunction with these, he also spoke of a 'primary' and a 'secondary' language. One soon finds out that in Wittgenstein's writings 'primary language' is just another name for the 'phenomenological language', while 'secondary language' is another name for the 'ordinary, physical language' (*PR*, § 157).[19] So one can establish two equivalences:

secondary system/language	↔	physical language
primary system/language	↔	phenomenological language

One should add to these equivalences Wittgenstein's distinction between *Hypothesen* and *Aussagen*. This distinction plays a central role in the manuscripts of 1929 and all indications are that it is parallel to the first two. Therefore one has the following situation:

secondary	↔	physical	↔	Hypothesis
primary	↔	phenomenological	↔	Assertion

Wittgenstein summed up his new conception in a conversation with his student Desmond Lee:

A hypothesis goes beyond immediate experience.
A proposition does not.
Propositions are true or false.
Hypotheses work or don't work.
A hypothesis is a law for constructing propositions, and the propositions are instances of this law. If they are true (verified), the hypothesis works; if they are not true, the hypothesis does not work. Or we may say that a hypothesis constructs expectations which are expressed in propositions and can be verified or falsified. (*LWL*, p. 110)

This is, in the briefest possible terms, Wittgenstein's sketch of his 'phenomenology'. The first or 'primary' level of language would be that of phenomenological assertions. In a nutshell, these are, not unlike elementary propositions in the *Tractatus Logico-Philosophicus*, descriptions of immediate experience. When Wittgenstein spoke in *Philosophical Remarks* of the need for a phenomenological colour theory as opposed to a physical one, he wrote that the former ought to be 'a theory in *pure* phenomenology in

[19] See also *PR*, § 71.

which mention is only made of what is actually perceptible and in which no hypothetical objects—waves, rods, cones, and all that—occur' (*PR*, § 218). Such remarks indicate clearly that Wittgenstein had in mind at the time a theory whose assertions refer to immediate experience without making any use of physical concepts, i.e. without any reference to 'objects' such as waves, cones, and rods. The phenomenological or 'primary' statements, as he called them, were meant to 'treat of what is immediate' (*PR* § 11). Presumably, they would be directly verified or refuted by our immediate experience, therefore either true or false. As Alva Noë put it, these assertions would be like 'a transparency through which immediate experience can be examined' (1994: 24). At this point Wittgenstein even toys with the notion of 'sense-data'. For example, he said in his Cambridge lectures that '[a] proposition is a judgement about sense-data, a reading of one's sense-data; for example "This is red" ' (*LWL*, p. 66). But a phenomenalist reading of Wittgenstein's primary language is not the only possibility.[20] To come back briefly to the debate surrounding the interpretation of the *Tractatus Logico-Philosophicus*, Waismann wrote in his *Theses* that the 'propositions that deal with reality immediately are called *elementary propositions*' (*WVC*, p. 248) and that 'Phenomena (experiences) are what elementary propositions describe' (*WVC*, p. 249). From such statements one could infer, as Merrill and Jaakko Hintikka did, that the elementary propositions of the *Tractatus Logico-Philosophicus* are propositions of a

[20] One might object that Wittgenstein's position, as construed here, boils down to phenomenalism and that it is unlikely, e.g. by virtue of reasons set forth in Pears (1987), that Wittgenstein was a phenomenalist. This is a very difficult question which cannot be dealt with properly here, but I can say a few words in defence of my interpretation. To begin with, I quoted Wittgenstein extensively on purpose; if there is any ambiguity, it will be found in his own remarks. At any rate, to say that phenomenological assertions 'treat of the immediate' does not necessarily mean that they are about sense-data, although Wittgenstein sometimes speaks as if they are, e.g. when he says that '[t]he world we live in is the world of sense-data; but the world we talk about is the world of physical objects' (*LWL*, p. 82). Experience gives us, according to Wittgenstein in the *TLP*, the objects out of which the world is built and not mere impressions of them. Therefore, one should not confuse this view with any form of Berkeleyian 'phenomenalism' according to which objects are constructs from sense-data. Merrill and Jaakko Hintikka never claimed that Wittgenstein held a form of 'phenomenalism'. This is a widespread misinterpretation, e.g. in Pears (1987: 65) and Carruthers (1990: ch. 8). Cf. Hintikka and Hintikka (1986: 72–4), Hintikka (1990: 157), and Hintikka (1993: 29–31). Wittgenstein's remarks on the notion of sense-data certainly deserve closer scrutiny, but such an investigation lies outside the scope of this book.

phenomenological language.[21] But one must remain cautious in making such a connection, since much of the structure of the *Tractatus Logico-Philosophicus* was brought down by the removal of the thesis of the logical independence of elementary propositions. It is therefore not clear whether Wittgenstein was really speaking about the same thing or not when he spoke of elementary propositions in the *Tractatus Logico-Philosophicus* and in 1929.

The second level would be that of 'physical' hypotheses. While phenomenological assertions are verified or falsified directly by immediate experience, hypotheses are described by Wittgenstein as *Gesetze zur Bildung von Sätzen* (*PR*, § 228) or *Gesetze zur Bildung von Aussagen* (*WVC*, pp. 99, 255), i.e. as 'laws for the formation of assertions'.[22] What is peculiar to hypotheses is that they are, in his own words, 'coupled with reality—with varying degrees of freedom' (*PR*, § 225). There seem to be only two cases envisaged by Wittgenstein: on the one hand, there is the limit case where reality will never conflict with the hypothesis. In that case it is *sinnlos* (*WA* ii, p. 113; *PR*, § 225). On the other hand, hypotheses are also described as *Gesetze zur Bildung von Erwartungen* or 'laws for forming expectations' (*PR*, § 228)[23] and, as in the case of physical laws, admit of an open-ended number of future confirmations. Since there can be no complete verification of the potentially infinite number of statements generated by the hypothesis, it can never be said to be definitely verified, and this means, according to Wittgenstein, that it has an altogether different relation to reality from that of an ordinary statement:

> When I say an hypothesis isn't definitively verifiable, that doesn't mean that there is a verification of it which we may approach ever more nearly, without ever reaching it. That is nonsense—of a kind into which we frequently lapse. No, an hypothesis simply has a different formal relation to reality from that of verification. (Hence, of course, the words 'true' and 'false' are also inapplicable here, or else have a different meaning.) (*PR*, § 228)

[21] See also *PG*, pp. 210–12 for an interesting critical discussion of the *TLP*. In this passage, Wittgenstein gives 'Here there is a red rose' as an example of elementary proposition. He also refers to his previous ideas: 'I used to think that . . . one would be able to use visual impressions etc. to define the concept say of a sphere, and thus exhibit once and for all the connections between concepts and lay bare the source of all misunderstandings, etc.' (p. 211).

[22] For more passages on the notion of a 'hypothesis' as a 'law for forming assertions', see *LWL*, p. 83 and *PG*, p. 219.

[23] Once again, Wittgenstein's use of the expression 'expectation' shows Ramsey's influence.

Clearly, Wittgenstein's 'hypotheses' are not assessable in terms of truth and falsity.[24] This notion of 'hypothesis' is thus remarkably similar to Ramsey's notion of 'variable hypothetical' described in section 4.1 above.[25] This connection becomes clearer in light of Wittgenstein's change of mind on quantification.

One important point to insist on is that the distinction between hypotheses and assertions had to be, in Wittgenstein's mind, as sharp as possible. For example, he insisted that 'the point of talking of sense-data and immediate experience is that we're after a description that has nothing hypothetical in it. If an hypothesis can't be definitely verified, it can't be verified at all, and there is no truth or falsity for it' (*PR*, § 226); and he told Schlick and Waismann, 'I must not produce any *hypotheses* concerning what is immediately given' (*WVC*, p. 97). It is worth noting in passing that Wittgenstein's definition of the phenomenogical as being free of everything hypothetical squares rather well with the use of the expression 'phenomenology' amongst physicists such as Mach or Boltzmann. These physicists used this expression as a label for the view that physics should 'represent nature without in any way going beyond experience' (Boltzmann 1974: 96), i.e. the view that it should avoid the use of purely theoretical—hypothetical—concepts.[26] It is also worth noticing the connection between this notion of 'hypothesis' and the conception of a

[24] The parallels between Wittgenstein's notion of 'hypothesis' and the notion used by German physicists such as Mach, Boltzmann, Hertz, and Weyl should not go unnoticed. The latter notion is very different from Newton's 'hypotheses' as encapsulated in his famous 'hypothesis non fingo'. Yvon Gauthier pointed out to me that Bernard Riemann is at the origin of this other, more Kantian, notion of 'hypothesis'. Riemann wrote in a posthumously published fragment: 'One must understand by hypothesis everything that thought adds to phenomena' (1953: 525), and in the same passage he gave a formulation of the law of inertia in the form of a conditional.

[25] One should notice here that Wittgenstein used in similar contexts the expression 'dictionary', e.g. in *M*, p. 297, which is typically Campbell's. As far as I know, there are no references to Campbell in the whole of Wittgenstein's *Nachlass*. Since Wittgenstein usually mentions the physicists (such as Boltzmann, Hertz, or Eddington) whose books he read, there is no reason to believe that he ever read Campbell's book. He may, however, have heard the expression in conversations with Frank Ramsey, who also used it in 1929 manuscripts such as 'Theories' (Ramsey 1978: 104).

[26] The expression 'phenomenology' is still in use in contemporary physics. Boltzmann presented 'mathematical phenomenology', in its 'most extreme form', as the view that 'physics must henceforth pursue the sole aim of writing down for each series of phenomena, without any hypothesis, model or mechanical explanation, equations from which the course of the phenomena can be quantitatively determined; so that the sole task of physics consisted in using trial and error to find the simplest equations that satisfied certain required formal conditions . . . and then to compare them with experience' (Boltzmann 1974: 95).

language as a *Satzsystem* which was prevalent in Wittgenstein's early transitional writings. As a result of the abandonment of the logical independence of elementary propositions, he defended time and again the view that propositions are always part of a *Saztsystem*.[27] Hypotheses put together form such a network or *Saztsystem*. Indeed, to take the example of equations of physics, which were thought by Wittgenstein to be hypotheses (*WVC*, pp. 159, 255), they form a system, producing new phenomenological statements which can in turn be verified: 'Physics constructs a system of hypotheses represented as a system of equations. The equations of physics can neither be true or false. It is only the findings in the course of a verification, i.e. phen[omenological] assertions, that are true or false' (*WVC*, p. 101). We saw in the previous section that Wittgenstein realized that there is a 'construction' in language which is not captured by truth-functional logic. I take it that hypotheses, from which one can derive new assertions, are precisely that which 'constructs' in language outside logic.

We saw that the distinction between 'primary' and 'secondary' systems originated in Hertz's *Principles of Mechanics*. But where does the distinction between 'hypotheses' and 'assertions' and their associated 'physical' and 'phenomenological' languages come from? There is a fundamental distinction in the *Tractatus Logico-Philosophicus* itself between *Sätze* and *Scheinsätze*. The latter were said to be lacking an important ingredient of statementhood, 'bipolarity', i.e. the ability to be true *and* to be false. This bipolarity was shown in the notation of the *Tractatus Logico-Philosophicus* by representing propositions with two poles, as in T*p*F. The truth-table for *p* & *q* looked like Fig. 1.

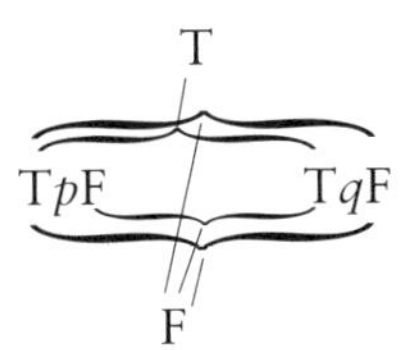

The class of *Scheinsätze* includes mathematical equations (*TLP*, 6.2). Logical propositions do not have truth-conditions; they are 'unconditionally true' (*TLP*, 4.461). In that peculiar sense and only in that sense they are not genuine propositions either,[28] although

[27] See e.g. *WVC*, pp. 63–4.

[28] See Dreben and Floyd (1991).

Wittgenstein does not consider them explicitly as *Scheinsätze*. The status of the laws of physics is also relatively unclear. There seems to be in the *Tractatus Logico-Philosophicus* a distinction between 'principles', such as the Law of Causality, and actual, specific laws of physics. The former are surely not tautologies (*TLP*, 6.31), but they are akin to logical propositions in that they can be said to be a priori. First of all, they assert nothing (*TLP*, 6.342), but show the possibility of propositions having a given form; they are, as Wittgenstein says, '*a priori* intuitions of possible forms of the propositions of science' (*TLP*, 6.34). Principles such as the Law of Causality are class names: 'the Law of Causality is not a law but the form of a law' (6.32). So 'in physics there are causal laws, laws of the causality form' (6.321). Moreover, we adopt them as part of the framework within which descriptions of the world take place: they provide 'the bricks for building the edifice of science' (6.341); they 'treat of the network and not of what the network describes' (6.35). In both respects, principles such as the Law of Causality are akin to *Scheinsätze*. But the causal laws are themselves propositions with sense, of a pictorial nature, and, although they are open-ended generalizations, they seem to admit a truth-functional treatment (quantifiers being infinite conjunctions or disjunctions).

Is the distinction between *Sätze* and *Scheinsätze* carried over to the 1929 manuscripts? Although it is undoubtedly the source of the above equivalences, there are some noticeable differences. First, although Wittgenstein did not specifically speak in 1929 of mathematical equations as *Scheinsätze*, his analysis of mathematics leads to similar conclusions: according to him, a mathematical theorem, if proved by induction, is not equivalent to a statement about 'all' numbers, because an *Induktion* does not assert anything (*WVC*, p. 82). Therefore such theorems are not assertions and they are not assessable in terms of truth and falsity. Secondly, causal laws acquired the status of 'hypothesis' and lost their ability to be assessable in terms of truth and falsity; they are no more assertions than mathematical theorems are.[29] The biggest change, however, is that propositions of ordinary language were by then also considered by Wittgenstein as hypotheses.[30] He said so explicitly in a 1931 lecture: ' "Propositions"

[29] It should be noted in passing that Griffin confused the view of causal laws in *TLP* with the new notion of 'hypothesis' in his book *Wittgenstein's Logical Atomism* (1964: 102–8). For a better discussion, see McGuinness (1971).

[30] This crucial move has gone mostly unnoticed in the secondary literature. One notable exception is Noë (1994: 13–14).

about physical objects and most of the things we talk about in ordinary life are always really hypotheses' (*LWL*, p. 53). When discussing the notion of hypothesis, Wittgenstein's examples are propositions of ordinary language such as 'This egg comes from a lark' (*WVC*, p. 101), 'This man is ill', 'The sun will rise tomorrow' (*LWL*, p. 66), 'There is a book lying there' (*PG*, p. 219), or 'There is a chair here' (*PG*, p. 220).

So, if the distinction between hypotheses and assertions is not really that of the *Tractatus Logico-Philosophicus* under a new guise, where does it come from? Part of the answer undoubtedly lies in Frank Ramsey's papers. I have just pointed out, however, that Wittgenstein went much further than Ramsey (or Hertz and Campbell for that matter) and boldly extended the notion of hypothesis to ordinary language. This generalization, which is worth a closer examination, should not come as a surprise if we reflect on the fact that ordinary language is 'physical' in Wittgenstein's sense: it achieves simplification by using names for objects, and no proposition making use of such names can count as a strictly phenomenological description.[31] The clue to this move is that Wittgenstein, having discarded his earlier notion of logical form, now sees the notion of object—the key notion of a physical language—as a hypothesis:

> When I say, 'All the different pictures I see belong to one object, say, to a table', that means that I connect the seen pictures by means of an hypothetically assumed law. On the basis of that law from given pictures I can derive a new picture. If I wanted to describe the particular aspects, that would be tremendously complicated. The structuring achieved by our language consists, therefore, in assembling all those innumerable aspects in a hypothetically assumed connection. . . . The language of everyday life uses a system of hypotheses. (*WVC*, p. 256)

In the same passage from Waismann's *Theses*, one finds the analogy between 'object' and a body in space:

> the particular aspects are the cross-sections made when we cut through it. What we observe are always only particular cross-sections across the connected structure of a law. If I know a number of cross-sections, I can connect them by means of an hypothesis. In the same way I can connect some aspects by means of an hypothesis. What connects them is nothing other than the object in question. (*WVC*, pp. 256–7)

[31] For the links between Wittgenstein's 'physical language' and Carnap's 'physicalist language' as defined in his 1931 paper, see Hintikka (1993).

In *Philosophical Remarks*, Wittgenstein wrote:

> All that is required of our propositions (about reality) to have a sense, is that our experience *in some sense or other* either tends to agree with them or tends not to agree with them. That is, immediate experience need confirm only something about them, *some* facet of them. And in fact this image is taken straight from reality, since we say 'There is a chair here', when we only see *one* side of it. (§ 225)

The above remarks are connected to Wittgenstein's rejection of Russell's notion of 'object'. His project, as it is clearly stated in 'The Relation of Sense-Data to Physics', had been to provide a 'logical construction' of physical objects out of sense-data (Russell 1986: 149–52). In *Our Knowledge of the External World*, Russell wrote that '(s)tarting from a world of helter-skelter sense-data, we wish to collect them into series, each of which can be regarded as consisting of the successive appearances of one "thing"' (Russell 1993: 107), and he defined a 'thing' as a 'series of aspects which obey the laws of physics' (1986: 165; 1993: 115–16). But Wittgenstein believed instead that 'an object is a connection of aspects represented by an hypothesis' (*WVC*, p. 256), and he thought that the notion involved in Russell's definition, which is that of 'class', does not provide us with the means to derive new assertions about new aspects, contrary to the notion of hypothesis:

> Russell does not represent the nature of objects correctly when he conceives of an object as a class. For a class does not help us at all to obtain a statement about a further aspect. A class has nothing to do with induction—an object, however, is essentially connected with induction. (*WVC*, p. 257)[32]

And the induction comes precisely in the form of a hypothesis! This is clearly stated earlier on in Waismann's *Thesen*:

> The truth of the matter is that the concept of an object is connected with *induction*. Induction appears in the form of hypotheses.
>
> By an *hypothesis* we here mean not a statement but rather a law for constructing statements. (*WVC*, p. 255)[33]

To say something about the respective merits of phenomenological and physical languages, Wittgenstein thought that in the end both

[32] Such remarks are echoes of the distinction between 'function' and 'operation' in *TLP*. On this distinction, see section 2.1 of Granger (1990).

[33] Here again, Wittgenstein was thinking about physics: 'There are hypotheses of mathematical form. The laws of physics are such hypotheses' (*WVC*, p. 255).

languages describe the same thing. He wrote, indeed, that 'the phenomenological language represents the same as our usual physical language' (*WA* i, p. 193), and that 'the physical language describes also only the primary world and not an hypothetical world' (*WA* i, p. 190). Now it is clear that Wittgenstein considered ordinary language as primarily a physical language. This does not mean, however, that ordinary language does not also possess the means necessary to describe immediate experience: 'our ordinary language is also phenomenological' (p. 4). It seems that Wittgenstein thought that ordinary language, being what one would call the universal medium (Hintikka and Hintikka 1986: ch. 1), possesses enough expressive power for one to be able to carve a purely phenomenological language from it. Wittgenstein also insisted frequently on the fact that a physical language has at least one important advantage over a phenomenological language, that of simplicity. Since phenomenological statements must reflect the multiplicity of the phenomena that they represent, their structure will invariably turn out to be extremely complex. Real numbers 'can appear in elementary propositions' (*WVC*, p. 42) but, according to Wittgenstein, ordinary language achieves great simplification by using terms for physical objects:

> Describing phenomena by means of the hypothesis of a world of material objects is unavoidable in view of its simplicity when compared with the unmanageably complicated phenomenological description. If I can see different discrete parts of a circle, it's perhaps impossible to give precise direct description of them, but the statement that they're parts of a circle . . . is simple. (*WA* i, p. 126; *PR*, § 230)

I would now like to draw attention to a few important consequences of the conceptions set forth so far. First, it should be clear by now that the relation between ordinary (physical) language and phenomenological language is that of hypothesis to assertion. The relation between hypothesis and assertion is fundamentally different from the relation between complex and elementary propositions to be found in the *Tractatus Logico-Philosophicus*, so Wittgenstein's notion of analysis in 1929 is accordingly a fundamentally different one.[34] With regard to what I said earlier about Wittgenstein's notion of elementary proposition in the *Tractatus Logico-Philosophicus* and

[34] As far as I know, this important consequence was first diagnosed by Noë (1994: 15).

in the phenomenology of 1929, it seems to me important not to confuse these notions.

There are many problems facing Wittgenstein's phenomenological inquiry. To begin with, the structure of phenomenological statements will presumably turn out to be extremely complex and, as he pointed out in 'Some Remarks on Logical Form', the task is 'very difficult' and philosophers have 'hardly yet begun to tackle it at some points' (p. 163). That the task is difficult is no reason to give it up, however, and reasons for the abandonment of the phenomenological project had to come from elsewhere. I take it that they came from the realization that a purely phenomenological language is simply not conceivable. Time and again Wittgenstein comes back in his manuscripts of 1929 to this question (he was, however, mostly preoccupied with questions in philosophy of mathematics), and he expressed his frustation by comparing phenomenological language with a 'magical swamp where everything conceivable disappears' (*WA* i, p. 192). He did not, however, abandon his line of inquiry until October 1929. Merrill and Jaakko Hintikka were the first to attract attention to striking evidence to that effect (*WA* ii, pp. 90–104). They originally presented Wittgenstein's 'deduction', as they called it, in the following terms:

> In briefest possible terms, Wittgenstein's 'deduction' thus ran as follows: the basic sentences of our language must be compared directly with (virtually, superimposed on) the facts they represent. But since language itself belongs to the physical world . . . such comparisons must take place in the physical world. Hence only what there is in the physical world can be represented directly in the language. (Hintikka and Hintikka 1986: 166)

For reasons of space, I shall confine myself to a brief sketch of this 'deduction'. Three elements are involved. First, there is a robust form of verificationism. Wittgenstein held the view that verification *is* the sense of a phenomenological assertion and that verification consists of a direct confrontation with reality. For example, he wrote just before October that '[y]ou cannot compare a picture with reality, unless you can set it against it as a yardstick. You must be able to fit the proposition on to reality' (*WA* ii, p. 89; *PR*, § 43). Secondly, there is the realization that language, phenomenological or not, belongs to the secondary, physical system: 'Language itself belongs to the second system. If I describe a language, I am essentially describing something that belongs to physics' (*WA* i, p. 191; *PR*, § 68). Realizing this, he immediately asks: 'But how can a physical language describe the phenomenal?' (*WA* i, p. 191; *PR*, § 68)

The third and last element of the 'deduction' is time. As mentioned earlier, Wittgenstein had come to the conclusion in 'Some Remarks on Logical Form' that phenomenological assertions contain numbers: 'real numbers or something similar to real numbers can appear in elementary propositions' (*WVC*, p. 42). It seems therefore that the process of verification of the statements will involve calculations, manipulations of symbols, which can only take place in physical time. Wittgenstein proposed the following (number-free) *Gedanken-experiment* in MS 106:

> Suppose I had such a good memory that I could remember all my sense impressions. In that case, there would, *prima facie*, be nothing to prevent me from describing them. This would be a biography. And why shouldn't I be able to leave everything hypothetical out of this description?
>
> I could e.g., represent the visual images plastically, perhaps with plaster-cast figures on reduced scale which I would only finish as far as I had actually seen them, designating the rest as inessential by shading or some other means.
>
> So far, everything would be fine. But what about the time I take to make this representation? I'm assuming I'd be able to keep pace with my memory in 'writing'—producing this representation. But if we suppose I then read the description through, isn't it now hypothetical after all? And why not? (*WA* i, p. 190; *PR*, § 67)

The upshot of this passage is that time is unavoidable in any description of immediate experience. Admitting thus that '[w]hat we understand by the word "language" unwinds in physical time' (*WA* i, p. 191; *PR*, § 69), Wittgenstein had driven a wedge between phenomena and language, and he was left pondering:

> Isn't it like this: a phenomenon (specious present) contains time, but isn't in time?
>
> Its form is time, but it has no place in time.
>
> Whereas language unwinds in time. (*WA* i, p. 191; *PR*, § 69)[35]

If the world of data is timeless, how can we speak of it at all? The stream of life, or the stream of the world, flows on and our propositions are so to speak verified only at instants. Our propositions are only verified by the present. (*PR*, § 48)

[35] The reference to 'specious present' is a clear indication of the Russellian background of Wittgenstein's cogitations. Russell had borrowed the notion from William James (James 1983: ch. 15), and made extensive use of it in his 1913 *Theory of Knowledge* (Russell 1984: ch. 6).

If language and the process of verification are recognized as part of the secondary, physical system, then it becomes obvious that (to paraphrase Wittgenstein) one cannot fit a proposition onto primary, phenomenological reality. It is only in the remarks dated 11 October that we find Wittgenstein realizing, for the first time, that he had reached a dead end:

> The immediate is in the grasp of a constant change. (It has in fact the form of a stream.) It is clear that if one wants to say here the last word, one will instead come to the limit of the language in which it is to be expressed. . . . The worst philosophical errors always arise when we apply our ordinary—physical—language in the area of the immediately given. . . . All our forms of speech have been taken from the ordinary, physical language and cannot to be used in the theory of knowledge or phenomenology without casting a distorting light on the object. The very expression 'I can perceive *x*' is itself taken from the idioms of physics, and *x* ought to be a physical object—e.g. a body—here. Things have already gone wrong if this expression is used in phenomenology, where *x* must refer to a datum. For then 'I' and 'perceive' also cannot have their previous senses. (*WA* ii, pp. 86–7; in part, *PR*, § 57)

In this passage Wittgenstein was, for the first time in his manuscripts, asserting the priority of physical over the phenomenological language: '[a]ll our forms of speech have been taken from the ordinary, physical language.' He had realized that in between the physical language and the very limits of language there is no room for a phenomenological language which would 'express what we really know' (*WVC*, p. 45). The 'most immediate description we can possibly imagine' (*WA* i, p. 191; *PR*, § 68) is already in the physical language. To use an earlier passage, what would otherwise come out would be 'that inarticulate sound with which many writers would like to begin philosophy' (*WA* i, p. 191; *PR*, § 68). To this remark Wittgenstein immediately added this condemnation: 'You simply can't begin before the beginning' (*WA* i, p. 191; *PR*, § 68). (These remarks occur much earlier, at the end of MS 106, and they express Wittgenstein's frustration with the notion of a phenomenological language. He had not found his argument yet.) After a few more days reconsidering his views, Wittgenstein reached the conclusion, on 22 October, that the very idea of a phenomenological language is absurd: 'The assumption that a phenomenological language is possible and that only it would say what we must express in philosophy is—I believe—absurd. We must get along with ordinary language and merely understand it better' (*WA* ii, p. 102). So, in October 1929, Wittgenstein realized that he could not conceive a purely phenomenological language of the sort needed for his phenomenology. This was the beginning of the end for this very project. Many years later, Wittgenstein was to write in *Remarks on Colour*:

> But what kind of a proposition is that, that blending in white removes the colouredness from the colour?
>
> As I mean it, it can't be a proposition of physics.
>
> Here the temptation to believe in a phenomenology, something midway between science and logic, is very great. (*RC*, ii, § 3)

To my mind, it is fairly obvious that Wittgenstein is referring here to his erstwhile temptation to set up a phenomenology, which he finally overcame in 1929, partially as a result of the 'deduction' of the impossibility of a purely phenomenological language. He never denied afterwards that there were 'phenomenological' problems;[36] it was only the idea that setting up a phenomenological language would help settling them which he had given up.

One of the obvious consequences of the abandonment of a phenomenological language is the transformation of the phenomenology, previously understood as a 'logical analysis of the phenomena themselves', into the fully-fledged philosophical 'grammar' of the mid-1930s onwards. As I have said, the change of mind meant that the project of a 'logical investigation' and a fortiori the very project of constructing an adequate logical notation (held by Frege, Russell, and Wittgenstein himself up to this point) had to be given up,[37] since not only did it rely on a misguided view of ordinary language as concealing what is essential to the method of representation, it was also based, in his own case, on the seemingly implausible idea of a purely phenomenological language. Wittgenstein had to abandon the very idea of a 'logical investigation'. He had to abandon the search for the logical form(s) of elementary (phenomenological) propositions; there was none waiting to be 'discovered'.

Since the construction of an adequate logical notation was abandoned, Wittgenstein thought that the best way to proceed in philosophy was to compare already existing different methods of representation—this is an important aspect of the later Wittgenstein

[36] Wittgenstein stated in *Remarks on Colour* that 'there is no phenomenology, but indeed there are phenomenological problems' (*RC* i, § 53). In fact, he tackles many typical 'phenomenological' problems in *Remarks on Colour*, e.g. i, §§ 59 f.

[37] On this point, I am in full agreement with Noë (1994: 17).

philosophical method. He thought that with such comparisons one would obtain the same results as with the construction of a phenomenological language. This is precisely what he told Schlick and Waismann in December 1929:

> I used to believe that there was the everyday language that we all usually spoke and a primary language that expressed what we really knew, namely phenomena. I also spoke of a first system and a second system. Now I wish to explain why I do not adhere to that conception any more. I think that essentially we have only one language, and that is our everyday language. We need not invent a new language or contruct a new symbolism, but our everyday language is *the* language, provided we rid it of the obscurities that lie hidden in it. Our language is completely in order, as long as we are clear about what it symbolizes. Languages other than the ordinary ones are also valuable in so far as they show us what they have in common. (*WVC*, pp. 45–6)

And the opening section of the *Philosophical Remarks* reads:

> I do not now have phenomenological language, or 'primary language' as I used to call it, in mind as a goal. I no longer hold it to be necessary. All that is possible and necessary is to separate what is essential from what is inessential in *our* language. . . . if we so to speak describe the class of languages which serve their purpose, then in so doing we have shown what is essential to them and given an immediate representation of immediate experience. Each time I say that, instead of such and such a representation, you could also use this other one, we take further steps towards the goal of grasping the essence of what is represented. A recognition of what is essential and what is inessential in our language if it is to represent, a recognition of which parts of our language are wheels turning idly, amounts to the construction of a phenomenological language. (§ 1)

Early in 1929, Wittgenstein had already described phenomenology as 'the grammar of the description of those facts on which physics builds its theories' (*WA* i, p. 5), a description which he used again in the first section of the *Philosophical Remarks* (§ 1). As a result of his change of mind, however, Wittgenstein's conception of grammar had to change. At the time of writing the *Big Typescript* in 1933 (from which the *Philosophical Grammar* was culled), Wittgenstein focused on the idea that the rules of grammar are *arbitrary* and *autonomous* in the sense that they determine meaning and therefore cannot be responsible to any prior meaning or reality. In that sense, they are autonomous.[38] One must ask here: where did the idea of the auton-

[38] This is perhaps the central idea of the *Big Typescript*, but it is already found in his lectures in 1931 (*LWL*, pp. 43–4, 56–7, 111).

omy of grammar come from? Surely, if the above reading is correct, it did not come from either the *Tractatus Logico-Philosophicus* or from the projected phenomenology of 1929. Indeed, the structure of the phenomenological is to be found by an investigation of phenomena. It is by definition dependent on experience; it is anything but arbitrary. One can see, however, the arbitrariness of grammar as a consequence of Wittgenstein's change of mind of October 1929: he must have realized at one point that the impossibility of phenomenological language and the abandonment of the project of phenomenology implied that there is no such thing as a language whose grammar is determined by immediate experience.[39]

A brief digression on the subject of dogmatism. One of the important aspects of the later Wittgenstein's conception of the nature of philosophy is the thesis that in philosophy you cannot discover anything new (*WVC*, p. 183; *PI*, § 89). Part of the revolution initiated in the autumn of 1929 consisted precisely in the realization that he was wrong to think that, although one cannot foresee the logical form of elementary propositions, it will nevertheless be discovered after a 'logical investigation of the phenomena themselves'. There is a clear statement to this effect in a discussion with Schlick and Waismann in December 1931:

> I used to believe, for example, that it is the task of logical analysis to discover the elementary propositions. I wrote, We are unable to specify the form of elementary propositions, and that was quite correct too. It was clear to me that here at any rate there are no hypotheses and that regarding these questions we cannot proceed by assuming from the very beginning, as Carnap does, that the elementary propositions consist of two-place relations, etc. Yet I did think that the elementary propositions could be specified at a later date. Only in recent years have I broken away from that mistake. (*WVC*, p. 182)

Again, this is another aspect of the later Wittgenstein's thought which takes its final form only after the change of mind of October 1929.

If Wittgenstein indeed realized in October 1929 that a phenomenological language is either impossible or simply not necessary, and if he was, as a result of this realization, forced to give up the phenomenological language, what happened then to the distinction

[39] I owe this neat way of putting it to Alva Noë, who presents this reasoning as an elaboration of Merrill and Jaakko Hintikka's interpretation, which he then proceeds to criticize (Noë 1994: 24–5).

between hypothesis and assertion? Did he not have to give up this distinction? The fact is that he did not give it up, at least not immediately. In March 1930 he was back in Vienna, presenting his new ideas to Schlick and Waismann. He introduced on this occasion his notion of 'hypothesis' (*WVC*, pp. 99–100). As a result of such discussions, Wittgenstein's notion of 'hypothesis' was to have an important influence on the Vienna Circle—he accused Carnap of having stolen the idea from him (Hintikka 1993). Its influence also extended to postwar British philosophy, where we find a not-so-distant relative in Waismann's 'open-textured' concepts (Waismann 1968: 39–66, 91–121). There are also many resemblances between Wittgenstein's 'hypotheses' and Gilbert Ryle's 'inference-tickets' (Ryle 1990: 234–49), but the origin of Ryle's well-known notion and his denial that indicative conditionals admit assessment as true or false should be traced back to John Cook Wilson's claim that, all statements being categorical, hypotheticals could not really be statements (Cook Wilson 1926: 525–52).

If the distinction was not abandoned, then its meaning must have changed. One may find a confirmation of this fact in a passage from the *Philosophical Grammar* (its origin in § 32 of the *Big Typescript*) where Wittgenstein used an old simile (see *WVC*, pp. 256–7) and compared a hypothesis to a multi-faceted body: 'The best comparison for every hypothesis—something that is itself an example of an hypothesis—is a body in relation to a systematic series of views of it from different angles' (*PG*, p. 221). According to this analogy, in order to verify the hypothesis one of its faces is laid against reality and, when the hypothesis is *used* in such a way, it becomes a genuine statement: 'The hypothesis, if *that* face of it is laid against reality, becomes a statement' (*PG*, p. 221). This is an important development: the sharp distinction between hypothesis and assertion of 1929 has disappeared. Wittgenstein himself says: 'At all events, there can't be any distinction between an hypothesis used as an expression of an immediate experience and a proposition in the stricter sense' (*PG*, p. 221). The distinction is now based on the different uses of what is recognizably the same proposition. According to Wittgenstein, it is our use which turns a given proposition either into an assertion about immediate experience or into a hypothesis. This change did not appear as the result of a slow process, but came quite rapidly after the end of 1929. It was already discussed in his lectures of 1931–2:

Just as the same expression can be a proposition or an hypothesis, so the same expression can be an equation or an hypothesis. Unless we distinguish confusions occur. . . . the same expression can be an hypothesis or a proposition in the strict sense. So also a proposition can be a priori or empirical. (*LWL*, pp. 76–7)[40]

(The example discussed by Wittgenstein at that point is an elementary arithmetical equation.)

The notion of hypothesis was to disappear shortly after completing the *Big Typescript*, at the time Wittgenstein began dictating the *Blue Book* in 1933, and the original distinction was to be ultimately replaced by the well-known distinction between 'norms of description' and 'empirical statements'. A hypothesis is still not a grammatical rule, since it still describes experience, while a grammatical rule must not do so by definition. One finds in the *Yellow Book* (1933–4) a further distinction between 'the use of a sentence as a hypothesis and as a grammatical rule' (*AWL*, p. 71):

Suppose that a planet which to a certain hypothesis describes an ellipse does in fact not do so. We should then say that there must be another planet, unseen, acting on it. It is arbitrary whether we say our laws of orbit are right, or that they are wrong. Here we have a transition between an hypothesis and a grammatical rule. If we say that whatever observations we make there is a planet nearby, we are laying this down as a rule of grammar; it describes no experience. (p. 70)

I cannot stress enough the importance for Wittgenstein's later philosophy of the transformation of the hypothesis/assertion distinction into a distinction between the use of a sentence as a rule and its use as a statement. For example, it plays a major role in many of the later Wittgenstein's arguments, for example in the diagnostic of the Cartesian mistake in the private-language argument (*PI*, § 151). In *On Certainty*, a prominent role is given in argumentation to 'hinge' propositions. These are conceived as a sort of hinge around which debates about the status of other propositions can take place; because of their special status they cannot be open to doubt. Again, these 'hinge' propositions may recognizably be the same as empirical statements, but they possess a different status as a result of their use as 'hinges'. I shall simply give here two quotations:

we are interested in the fact that about certain empirical propositions no doubts can exist if making judgements is to be possible at all. Or again: I am

[40] See also *LWL*, pp. 110, 81.

inclined to believe that not everything that has the form of an empirical propositions *is* one. (*OC*, § 308)

It is clear that our empirical propositions do not all have the same status, since one can lay down such a proposition and turn it from an empirical proposition into a norm of description. (§ 167)

The distinction is also present in the first part of the *Remarks on Colour* (1951):

Sentences are often used on the borderline between logic and the empirical, so that their meaning changes back and forth and they count now as expressions of norms, now as expressions of experience.

For it is certainly not an accompanying mental phenomenon—this is how we imagine 'thoughts'—but the use, which distinguishes the logical proposition from the empirical one. (*RC*, i, § 32)

We are definitively in the presence here of one of the later Wittgenstein's major conceptual tools. This shows the centrality of the topics discussed in this chapter and the previous one for any global interpretation of Wittgenstein's philosophy. The fact is that these questions were usually given insufficient treatment in the past, and I hope that the foregoing remarks will serve as a sound basis for further research. I would like to turn now to a series of topics, such as the Law of Excluded Middle, mathematical existence, infinity and the continuum, where Wittgenstein's remarks are, once more, best understood in terms of a commitment to finitism.

6

Philosophy and Logical Foundations

6.1. INTUITIONISM, INTENTIONALITY, RULES, AND DECISION PROCEDURES

It is now time to have a closer look at some of the parallels and differences between intuitionism and Wittgenstein's 'new logic'. One good way to handle this topic is with the help of the notion of 'truth-maker'. This expression was introduced by Kevin Mulligan, Peter Simons, and Barry Smith in order to designate in a neutral fashion 'entities *in virtue of which* sentences and/or propositions are true' (Mulligan et al. 1984: 287). As is well known, Wittgenstein had in the *Tractatus Logico-Philosophicus* assigned to *Sachverhalten* or atomic facts the role of making propositions true.[1] The basic intuitionist thesis is, on the other hand, that a (mathematical) proposition is made true by a proof of it. It is on the basis of this thesis that intuitionists developed a new interpretation of logical constants in the 1930s which has interesting affinities not with conceptions found in the *Tractatus Logico-Philosophicus* but with Wittgenstein's new form of analysis in the early 1930s.

With the notable exception of Hermann Weyl's remarks on the quantifiers in his 1921 paper 'Über die neue Grundlagenkrise der Mathematik' (which in any case do not reflect adequately Brouwer's ideas), there was no attempt at developing a properly intuitionistic interpretation of the logical constants until the early 1930s. At that point both Arend Heyting, a student of Brouwer, and Nicolaï Kolmogorov, the well-known Russian mathematician who was to contribute significantly to modern probability theory, developed an intuitionistic sentential calculus based on two different but in the end equivalent readings of the logical constants. In terms of truth-maker analysis, the basic intuitionistic thesis is that

[1] For an introduction to the 'truth-maker' terminology and a fine analysis of the notion of *Sachverhalt* in *TLP*, see Mulligan et al. (1984) and Mulligan (1985). See also Sundholm (1994).

p is true = there exists a proof of p

The notion of truth-bearer which has been since adopted by intuitionists is that of an *Aussage*. Heyting himself distinguished between *Aussage* and *Satz* (1983: 113). For him a *Satz* is 'the affirmation of an *Aussage*'; it corresponds to what appears on the right of Frege's assertion sign ⊢, i.e. it is a theorem. Heyting's choice of words, without being in any way incorrect, might confuse some readers: if we think of a proposition as the content of what is propounded—as opposed to that which is propounded—then we can translate *Aussage* by 'proposition' and *Satz* by 'assertion' (an assertion being simply an asserted proposition). But the appropriate and usual translation of the German terms is precisely the reverse, i.e. 'assertion' for *Aussage* and 'proposition' for *Satz*. It is simply that Heyting uses *Satz* in its traditional German mathematical sense of 'theorem' or 'judgment proved'. This corresponds to the old English use of 'proposition' as 'that which is propounded', as opposed to the modern use followed here. Heyting's use of *Aussage*, which is close to Frege's use of *Gedanke*, is perhaps less fortunate, since 'declaration' is a more literal translation than 'proposition'.[2] In order to avoid any confusion, I shall use 'proposition' (*Satz*) where Heyting uses 'assertion' (*Aussage*) and vice versa. So when Heyting says that a *Satz* is the affirmation of an *Aussage*, I would say that an assertion (or *statement*) is the affirmation of a proposition.

Heyting made in the same text the further claim that a proposition expresses a certain 'expectation' (*Erwartung*). Heyting, who knew the work of Husserl and his follower Oskar Becker, also spoke of a proposition as being the expression of an 'intention' (*Intention*). Heyting's example is that of Euler's constant C. The proposition 'Euler's constant C is rational' expresses, according to him, the expectation or intention that we could find two integers a and b such that $C = a/b$. In terms of truth-maker analysis, what renders an expectation or intention true is its fulfilment, which is in turn affirmed in the assertion. Indeed, the affirmation of the proposition (i.e. the assertion) 'means the fulfilment of an intention'. In our example, it is the fact that we have finally found the integers a and b such that $C = a/b$. An assertion is not, however, a proposition but the determination (*Feststellung*) of the fact that the intention expressed in the proposition is fulfilled: 'The affirmation of a [proposition] is not itself a

[2] I am grateful to Göran Sundholm for clarifying these terminological matters.

[proposition]; it is the determination of an empirical fact, viz., the fulfilment of the intention expressed by the [proposition]' (Heyting 1983: 59). It is hard to understand exactly what Heyting meant. Surely, according to him:

An expectation (*Erwartung*) corresponds to a fulfilment (*Erfüllung*)

On this point, Kolmogorov's conception is perfectly similar.[3] Heyting seems to be construing the mathematical theorem expressing the rationality of Euler's constant as a statement rendered true by an empirical fact, namely the fact that a mathematician has found two integers such that $C = a/b$. This much must not be entirely right, since the claim is that 'the affirmation of a [proposition] is not itself a [proposition]'; in other words the theorem is not really a statement, but a determination (*Feststellung*) of a fact. One is reminded here of Wittgenstein, for whom, in speaking about natural numbers, one ought to use an inductive proof (an *Induktion*) which is neither an *Aussage* nor a *Satz*; the law of associativity is not asserting a proposition about all numbers, it is an *Induktion*, a template, which, strictly speaking 'asserts' nothing. But it is not exactly the same for Heyting. To take once more the example of the rationality of Euler's constant, it seems at any rate that according to Heyting there are not two but three terms: first, there is the expectation (or intention) of finding two integers such that $C = a/b$. Secondly, there is the fulfilment, i.e. the finding of the integers a and b. Finally, there is the 'affirmation' or 'determination' of the fact that the expectation has been fulfilled, in the theorem expressing the rationality of Euler's constant.

It is not clear to me what Heyting meant when he said that 'the affirmation of a [proposition] is not itself a [proposition]', i.e. that a theorem is not truly a statement but a 'determination' of a fact. As I have just pointed out, one is reminded here of Wittgenstein's idea that arithmetical theorems are not assertions (*Aussagen*) about 'all' natural numbers but, strictly speaking, inductions or templates, which assert nothing. Whether or not this corresponds to something

[3] In Kolmogorov's *Aufgabenrechnung* of 1932, the truth-bearer was not a proposition but a 'problem' or 'task' (*Aufgabe*) (Kolmogorov 1932: 58). To be more exact, Kolmogorov's position was that a proposition expresses a problem (as opposed to an expectation or intention) and a problem corresponds to a solution (*Lösung*). This truth-maker analysis is in fact equivalent to Heyting's. (This fact was recognized by Heyting only after the war, in Heyting 1958: 107.)

in Heyting's analysis, one should emphasize here that Wittgenstein did not consider general arithmetical propositions a 'logical product of equations' but instead as a 'law for the construction of equations' (*PG*, p. 432) and that he thought that precisely because they are not logical products, such 'laws' are not assessable in terms of truth and falsity. The closely related notion of hypothesis has a similar analysis: it is a law for the formation of propositions or expectations, and a proposition corresponds to a state of affairs while an expectation is fulfilled. Every proposition or expectation generated by the hypothesis is verifiable, but the hypothesis itself is not definitively verifiable; hence it is not strictly speaking true or false. Thus propositions and expectations are truth-bearers and the calculus of truth-functions applies to them, but there is no room in Wittgenstein's conception for a renovated interpretation of logical connectives which would legitimate their use where an arithmetical induction or a hypothesis is involved.

With respect to the notions of intention, expectation, and fulfilment, there are affinities between Wittgenstein and intuitionism which will be readily seen in the use of a common terminology.[4] There is, however, an important difference. In order to discuss it, I must introduce a digression about the role of intention in language according to Wittgenstein in the early 1930s, beginning by recalling a crucial change to the picture theory. The conception of propositions as pictures of the *Sachverhalten* is surreptitiously replaced in *Philosophical Remarks* by a much less static one which is a forerunner of the later account of meaning as use. The transition is obvious in passages such as this: 'If you think of propositions as instructions for making models, their pictorial nature becomes even clearer' (*PR*, § 10). The notion of logical or mathematical multiplicity still plays a role, but the earlier requirement that the proposition must possess the same multiplicity as the *Sachverhalt* it is picturing (TLP, 4.04) is transformed into the requirement that language must possess the same multiplicity as the actions that it sets off:

> Language must have the same multiplicity as a control panel that sets off the actions corresponding to its propositions . . . Just as handles in a control room are used to do a wide variety of things, so are the words of language that corresponds to the handles. . . .
>
> A word only has meaning in the context of a proposition: that is like saying only in use is a rod a lever. Only the application makes it into a lever.

[4] These affinities had already been noticed by Sundholm (1994: 121).

Every instruction can be construed as a description, every description as an instruction. (*PR*, §§ 13–14)

In a conception of this kind intentionality plays a key role since, to put it crudely, one has to *intend* to use the instruction-description in a certain way. Wittgenstein knew the importance of intentionality: 'If you exclude the element of intention from language, its whole function then collapses' (*PR*, § 20). The same idea is expressed in a lecture of 1930, in terms of commitment:

if a proposition is to have any sense we must commit ourselves to the use of the words in it. It is not a matter of association; that would not make language work at all. What is essential is that in using the word I commit myself to a rule of use. (*LWL*, p. 36)

At that stage, Wittgenstein was conceiving intentionality as some kind of mental process. In *Philosophical Remarks* he wrote that '[w]hat is essential to intention is the picture: the picture of what is intended' (§ 21), adding later:

How is a picture meant? The intention never resides in the picture itself, since, no matter how the picture is formed, it can always be meant in different ways. But that doesn't mean that the way the picture is meant only emerges when it elicits a certain reaction, for the intention is already expressed in the way I *now* compare the picture with reality. (§ 24)

Wittgenstein is speaking here of intention as being given in the comparison between the picture—which is a mental entity—and reality. As noted by Jaakko Hintikka in his paper 'Rules, Games and Experiences: Wittgenstein's Discussion of Rule-Following in Light of his Development', such a comparison cannot be anything else than a mental process, it has a typically 'phenomenological' character in the sense considered in Chapter 5 (Hintikka 1989: 282).[5]

To come back to Heyting's analysis, Wittgenstein never commented on it. It has, however, a perfect counterpart in Russell's analysis of desire and intention in *The Analysis of Mind* that Wittgenstein rejected. In virtue of the strong parallels between Heyting and Russell on intention, Wittgenstein's criticisms of the latter will give a good indication of his stance on the former. In *The*

[5] It should be pointed out that the momentary character of understanding, alluded to in the passage from *PR*, § 24 just quoted, is one of the main features of the phenomenological conception of language and rules which was to be criticized at length in *PI*, §§ 139–42, 151–2, 155, 184, 197–8.

Analysis of Mind, Russell proposed an analysis of intentional acts in terms of 'external' relations between a proposition expressing an expectation or desire, the fact that constitutes the fulfilment of this expectation, and the recognition of the fulfilment or state of 'quiescence' (Russell 1989: 75). Since he characterized the object of desire as that thing the possession of which would end the state of desire, Russell was thus presenting what is in fact a dispositionalist account of desire and intentional acts in general.

There is an analysis of intentionality in terms of 'expectation' and 'fulfilment' in Wittgenstein's writings in the early 1930s (e.g. *LWL*, pp. 30–6, 44; *PR*, §§ 28, 33, 34); it even found its way into the *Philosophical Investigations* (§§ 442–65). For Wittgenstein the relation between an expectation and its fulfilment is not an 'external' but an 'internal' one (in the sense of the *Tractatus Logico-Philosophicus*), and the recognition of the fulfilment of an expectation is nothing more than the seeing of that internal relation, in 'our looking at the two terms and seeing the internal relation between them' (*LWL*, p. 57). Naturally, Wittgenstein disagreed with Russell's analysis by claiming that his conception

> regards recognition as seeing an internal relation, whereas in (Russell's) view this is an external relation.
>
> That is to say, for me, there are only two things involved in the fact that a thought is true, i.e. the thought and the fact; whereas for Russell, there are three, i.e. thought, fact and a third event which, if it occurs, is just recognition. (*PR*, § 21)

Wittgenstein had a simple, ironical argument:

> I believe Russell's theory amounts to the following: If I give someone an order and I am happy with what he then does, then he has carried out my order.
>
> (If I wanted to eat an apple, and someone punched me in the stomach, taking away my appetite, then it was this punch that I originally wanted.) (*PR*, § 22)

I shall make only two comments on this criticism of Russell's dispositionalist account of intention. First, although it has been discussed in secondary literature,[6] it seems that no one has seen that it is deficient in that Wittgenstein still relies, with his conception of the internal relation between an expectation and its fulfilment, on a

[6] See e.g. Gargani (1984); Shanker (1993).

phenomenological conception. Secondly, Russell's three-termed analysis of intention corresponds to the analysis by Heyting, presented above, in terms of expectation, fulfilment, and 'affirmation' in the theorem of the fact that the expectation has been fulfilled. Although Wittgenstein's argument cannot be made to apply to Heyting's analysis, it is a good indication that he would have rejected it.

With Wittgenstein's phenomenological conception of language and intentionality comes a phenomenological conception of rules. Indeed, it is natural to think a rule—any rule—needs to be (correctly) 'understood' or 'interpreted' in order to set off the right action. To see this simple fact, it suffices to take one example, such as the rule

$a_1 = a_2 = 1$
for all $n \geq 2$, $a_n = a_{n-1} + a_{n-2}$

which generates the Fibonacci sequence:

1, 1, 2, 3, 5, 8, 13, 21, 34, 55, 89, 144, 233, 377 . . .

If one looks at the expression of the rule as mere physical marks on paper, where does one get the instruction to obtain the next number of the sequence? Where is the link between the rule and the sequence of numbers it generates? It seems to follow that the symbolic expression of the rule cannot serve of itself as a guide, and thus that any action according to the rule must contain an interpretation of the rule. That is why Wittgenstein spoke of the rule as 'contained in our intention' (*LWL*, p. 40). He expressed the same idea by saying also:

> The method of projection must be contained in the process of projecting; the process of representation reaches up to what it represents by means of a rule of projection. If I copy anything the slips in my copy will be compensated for by my anger, regret, etc., at them. The total result—i.e. the copy *plus the intention*—is the equivalent of the original. The actual result—the mere visible copy—does not represent the whole process of copying; we must include the intention. The *process* contains the rule, the *result* is not enough to describe the process. (*LWL*, pp. 36–7)

This quotation gives us a clue to the source of the phenomenological conception of language in the *Tractatus Logico-Philosophicus*. It is to be found in 3.1–14, a passage whose interpretation is greatly helped by this quotation. According to Wittgenstein, the 'proposition is the propositional sign in its projective relation to the world' (*TLP*, 3.12), so a proposition has two components: the method of projection, which is 'the thinking of the sense of the proposition' (*TLP*, 3.11), and the propositional sign, which is 'a fact' (*TLP*, 3.14). Transposed in the terminology of the early 1930s, the propositional sign is a 'physical' entity, while the method of projection is a 'phenomenological' entity. In fact, 'method of projection' is here just another name for 'intention'. To my mind, there is no clearer evidence that the phenomenological conception was already in the *Tractatus Logico-Philosophicus*.

To repeat, Wittgenstein held the view that the symbolic expression of the rule cannot serve of itself as a guide; it needs to be interpreted, understood, intended, etc. But the interpreted rule is a phenomenological object, and such a conception should not have survived the abandonment of the phenomenological language presented in section 5.2 above: once Wittgenstein had realized that a proposition belonging to the physical world could neither be compared to nor be used to express a phenomenological fact, he should have realized that a rule conceived as a phenomenological object could neither be compared to nor be used to set off an action taking place in the physical world. Wittgenstein was facing a dilemma, since on the one hand the symbolic expression of the rule cannot serve of itself as a guide, and on the other the rule conceived as a phenomenological object could not be used to set off an action taking place in the physical world. I believe, along with Jaakko Hintikka, that this is the right diagnosis of Wittgenstein's on rules in the early 1930s, and that the remarks on rule-following in *Philosophical Investigations* (§§ 143–242) are the result of his efforts to find a way out of this dilemma (Hintikka 1989). I shall now show that Wittgenstein saw his new standpoint on rules as refuting the intuitionist understanding of rule-following. This is another important disagreement between him and intuitionism. The topic of rule-following will also play an important role in section 8.1 below, when I shall be criticizing the traditional interpretation of Wittgenstein as a strict finitist.

To see Wittgenstein's way out of this dilemma, one must first look at his efforts to work out a better physicalist account, and here one must trek back to 1931, when Wittgenstein began to describe 'intention' or 'understanding' as 'calculating' (*WVC*, p. 168) or 'operating' with words and propositions:

What does it mean to understand a proposition? This is connected with the general question of what it is what people call *intention, to mean, meaning*.

Nowadays the ordinary view is, isn't it, that understanding is a psychological process that accompanies the proposition—i.e. a spoken or written proposition? What structure, then, does this process have? The same, perhaps, as a proposition? Or is this process something amorphous, as when I read a proposition while I have a toothache?

I now believe that understanding is not a particular psychological process at all that there is in addition, supplementary to the perception of a propositional picture. It is true that various processes are going on inside me when I hear or read a proposition. An image emerges, say, there are associations and so forth. But all these processes are not what I am interested in here. I understand a proposition by *applying* it. Understanding is thus not a particular process; it is operating with a proposition. *The point of a proposition is that we should operate with it.* (What I do, too, is an operation.) (*WVC*, p. 167)[7]

One should notice the analogy between understanding and having a toothache. Wittgenstein comes back to it: 'The view that I wish to argue against in this context is that understanding is a *state* inside me, like, for instance, a toothache' (*WVC*, p. 167). This analogy with the phenomenological object *par excellence* shows clearly, if any evidence were needed, that it is against his phenomenological conception that Wittgenstein wishes now to argue. One may jump here to the *Blue Book*, where he was trying to make some room for the physical character of rules, when discussing the teaching of the meaning of the word 'yellow':

If we are taught the meaning of the word 'yellow' by being given some sort of ostensive definition (a rule of usage of the word) this teaching can be looked at in two different ways.

A. The teaching is a drill. . . . The drill causes us to associate a yellow image, yellow things, with the word 'yellow'. Thus when I gave the order 'Choose a yellow ball from this bag' the word 'yellow' might have brought up an image, or a feeling of recognition when the person's eye fell on the yellow ball. The drill of teaching could in this case be said to have built up a psychical mechanism. . . . the teaching is . . . the cause of the phenomena of understanding, obeying, etc. . . .

B. The teaching may have supplied us with a rule which is itself involved in the processes of understanding, obeying, etc.; 'involved', however, meaning that the expression of this rule forms part of these processes. (*BB*, pp. 12–13)

[7] Wittgenstein also says, a little further on: 'What I am doing with the words of a language in *understanding* them is exactly the same thing I do with a sign in the calculus: I operate with them' (*WVC*, pp. 169–70).

One finds once more both conceptions in this passage, since alternative A (teaching as a drill building up a mental mechanism) is no other than the now discredited phenomenological conception, and alternative B (teaching as providing an expression of the rule which is involved in the process of understanding) is nothing more than the physical conception. After using an arithmetical example, Wittgenstein came back to alternative B:

> We shall say that the rule is *involved* in the understanding, obeying, etc., if, as I should like to express it, the symbol of the rule forms part of the calculation. (As we are not interested in where the processes of thinking, calculating, take place, we can for our purposes imagine the calculations being done entirely on paper. We are not concerned with the difference: internal, external.) (*BB*, p. 13)

Clearly, Wittgenstein was trying to make space for his new physical conception by using the symbolic expression of the rule to replace the intention. But, again, the symbolic expression of the rule cannot by itself determine the way it is followed. As such it is not properly connected with the act: if the interpretation is thought of as another symbolic expression, it cannot act as a bridge between the rule and an act according to it. This point comes out particularly clearly in *Philosophical Investigations*, § 198:

> 'But how can a rule show me what I have to do at *this* point? Whatever I do is, on some interpretation, in accord with the rule'.—That is not what we ought to say, but rather every interpretation, together with what is being interpreted, hangs in the air; the former cannot give the latter any support. Interpretations by themselves do not determine meaning. (*PI*, § 198)

Wittgenstein condemned once and for all his (physical) conception of the symbolic expression of the rule at § 221 by saying that it was 'really a mythological description of the use of the rule' (§ 221). If alternative B appears now to be a dead end, can one infer that Wittgenstein was going to go for alternative A, i.e. for the 'teaching as a drill' solution? As Hintikka pointed out, it is only partly the case (Hintikka 1989: 194). At least the role of teaching is not conceived any more as a causal one. This is clear from Wittgenstein's discussion of signposts at the end of § 198:

> 'Then whatever I do be brought into accord with the rule?'—Let me ask this: what has the expression of a rule—say a sign-post—got to do with my actions? What sort of connexion is there here?—Well perhaps this one: I have

been trained to react to this sign in a particular way, and now I do so react to it.

But that is only to give a causal connexion; to tell how it has come about that we now go by the sign-post; not what this going-by-the-sign really consists in. On the contrary; I have further indicated that a person goes by a sign-post only in so far as there exists a regular use of sign-posts, a custom.

So now Wittgenstein's thought has shifted considerably: the criteria of rule-following no longer reside for him either in the (phenomenological) experience that one has when following it nor in the 'involvement' of the symbolic expression of the rule in the process of following it. It lies in the complex of customs of which an act of following a rule is a part. This is what Wittgenstein expressed by writing that 'a person goes by a sign-post only in so far as there exists a regular use of sign-posts, a custom'. Such complexes of customs are what he called 'language-games', and thus his point is that we can only speak of rule-following against the background of the language-game. As Jaakko Hintikka as shown, the conceptual priority of rules over games is reversed: rules are not constitutive of language-games; rather, it is the language-games that are prior to rules. This new conceptual priority is expressed thus: 'To understand a sentence means to understand a language. To understand a language means to master a technique' (*PI*, § 199).

One now sees how Wittgenstein could on the one hand reject his earlier phenomenological conception of rules while on the other hand answering the worry that led to its acceptance (i.e. the idea that the symbolic expression of the rule could not determine its applications of itself). The only criteria for determining whether a rule has been followed reside in the whole of the language-game. It is clear now that the symbolic expression of the rule no longer plays a role. It is not part of the criteria for following a rule, nor is it involved in the act. This is what Wittgenstein meant when writing: 'When I obey the rule I do not choose. I obey the rule blindly' (*PI*, § 219). Not that one does not use symbolic expressions when following rules; it is just that this is not what following the rule consists of. What it consists of is a move in a language-game. In the end, what 'struck' Kripke as being sceptical worries (Kripke 1982) is nothing more than an argument for the priority of the entire language-game over the symbolic expressions of its particular rules.

For the connection with intuitionism Brouwer spoke of a fundamental intuition (*Urintuition*), that of the duality of time (*Zweiheit*), upon which the natural number series is constructed. In his 1929 manuscripts, Wittgenstein—who was still faithful to his earlier view (presented in Chapter 2 above), stating that what is fundamental is 'the repetition of an operation' (*WA* i, p. 102)— described this 'fundamental intuition' incorrectly as the view that every application of a rule has its own individuality (*WA* i, pp. 101–2).[8] This conception of rule-following is eminently phenomenological. Indeed, by seemingly requiring a new 'intuition' of the rule at every step, intuitionists appeared to him as holding the very phenomenological conception of rules that Wittgenstein discarded. This is precisely the reason why Wittgenstein disagreed with them:

> Intuitionism comes to saying that you can make a new rule at each point. It requires that we have an intuition at each step in calculation, at each application of the rule; for how can we tell how a rule which has been used for fourteen steps applies at the fifteenth?—And they go on to say that the series of cardinal numbers is known to us by a ground-intuition—that is, we know at each step what the operation of adding 1 will give. We might as well say that we need, not an intuition at each step, but a decision.—Actually there is neither. You don't make a decision: you simply do a certain thing. It is a question of a certain practice.
>
> Intuitionism is all bosh—entirely. (*LFM*, p. 237)

To come back to the basic intuitionist thesis mentioned at the beginning of this section that a mathematical proposition is made true by a proof of it, there is one basic point shared by Wittgenstein: according to both, a proof is not a mere vehicle for recognizing the truth[9] but determines the sense of a mathematical statement or proposition (in what follows these expressions will be used interchangeably in order to keep in line with Wittgenstein's terminology).

[8] This mistake was pointed out to me by Michael Wrigley.

[9] Wittgenstein's contemporary in Cambridge, the mathematician G. H. Hardy, propounded a similar Platonist stance on proofs in his 1929 paper, 'Mathematical Proof' (Hardy 1929). Wittgenstein knew Hardy's paper, which he quoted frequently in his lectures (e.g. *AWL*, pp. 215–20, 222, 224–5 or, later, *LFM*, 91, 103, 123, 139, 169–71, 239, 243) and writings (one example of which is quoted in section 6.3), always criticizing it. For a careful analysis of Wittgenstein's criticisms of the 'Hardyian picture' of mathematics, see Gerrard (1987). Gerrard has propounded, on the basis of his 'Hardyian picture', one of the rare general overviews of the development of Wittgenstein's philosophy of mathematics, which takes into account writings of the transitional period (Gerrard 1991). However, his discussion (as well as that found in Gerrard 1990) focuses on Wittgenstein's remarks on contradiction in mathematics, a topic which is not dealt with in this book. For another *Übersicht* of Wittgenstein's philosophy of mathematics which makes place for writings of the transitional period, see Wrigley (1993).

Proofs by induction, which were discussed in section 4.2 above, provide us with a perfect example:

> Most people think that complete induction is merely a way of reaching a certain proposition; that the method of induction is supplemented by a particular inference saying, *therefore* this proposition applies to all numbers. Here I ask the question, What about this 'therefore'? There is no 'therefore' here! Complete induction *is* the proposition to be proved, it is the whole thing, not just the path taken by the proof. This method is not a vehicle for getting anywhere. In mathematics there are not, first, propositions that have sense by themselves and, second, a method to determine the truth or falsity of propositions; there is only a method, and what is called a proposition is only an abbreviated name for the method. (*WVC*, p. 33)

We find here Wittgenstein's idea, presented in section 4.2 above, that the result of a proof by induction is not adequately represented by an 'all' statement i.e. by a quantified statement of the form $\forall x\, F(x)$. As we also saw in section 5.2 above, Wittgenstein held a robust form of verificationism according to which verification (direct confrontation with reality) is the sense of a phenomenological assertion. This strong verificationism was extended to mathematics, where the method of verification—to be found in the proof—determines the sense of the statement. So in mathematics the content of a statement cannot be anything over and above what its proof shows:

> If you want to know what the expression 'continuity of a function' means, look at the proof of continuity; that will show what it proves. Don't look at the result as it is expressed in prose, or in the Russellian notation, which is simply a translation of the prose expression; but fix your attention on the calculation actually going on in the proof. The verbal expression of the allegedly proved proposition is in most cases misleading, because it conceals the real purport of the proof, which can be seen with full clarity in the proof itself. (*PG*, pp. 369–70)

It is worth noticing here Wittgenstein's use of his *Kalkul/Prosa* distinction (introduced in Chapter 1 above): the proof provides a method of calculation which determines the sense of the statement; prose is misleading, as one is led to believe that sense comes from somewhere else. (This point is in turn linked with Wittgenstein's criticisms of formalism and Frege's 'arithmetic with content' which will be presented in the next section.)[10]

[10] Wittgenstein was to speak later of the 'disastrous invasion' of mathematics by logic (*RFM* v, § 24). This controversial point is precisely linked with the idea that

It may be possible to talk of mathematical 'statements' or 'propositions', but the analogy has its limits since, Wittgenstein tells us, a mathematical proposition such as 'The equation $x^2 = 1$ has 2 roots' is of a *different kind* from a proposition outside mathematics such as 'There are 2 apples on the table' (*PG*, pp. 348, 350, 459). This difference in kind hangs, according to Wittgenstein, on the difference in the methods of verification, which consist in the first case of applying a procedure of decision (an algorithm) and, in the second, in a direct confrontation with reality. This point is put across in numerous passages. The following one contains, again, a direct reference to the ideas about generality and quantification presented in section 4.2 above: 'How a proposition is verified is what it says. Compare generality in arithmetic with the generality of non-arithmetical propositions. It is differently verified and so is of a different kind. The verification is not a mere token of the truth, but determines the sense of the proposition' (*PG*, pp. 458–9). This much is quite in line with new developments in Wittgenstein's thought: one should remember here from the beginning of section 5.1 above that already in January 1929 Wittgenstein had written that 'One could surely replace the logic of tautologies by a logic of equations' (*WA* i, p. 7). As late as 1933 he spoke in *Philosophical Grammar* about excluding 'propositions of arithmetic' from the scope of the 'general form of proposition' (*PG*, p. 124). In the same book he explored the possibility of an analogy between mathematical and ordinary propositions:

> A mathematical proposition that has been proved has a bias towards truth in its grammar. In order to understand the sense of $25 \times 25 = 625$ I may ask: how is this proposition proved? But I can't ask how its contradictory is or would be proved, because it makes no sense to speak of a proof of the contradictory of $25 \times 25 = 625$. So if I want to raise a question which won't depend on the truth of the proposition, I have to speak of *checking* its truth, not of proving or disproving it. (p. 366)

Thus the analogy between ordinary and mathematical propositions hangs on the notion of a 'method of checking':

> The method of checking corresponds to what one may call the sense of the mathematical proposition. The description of this method is a general one

focusing on the Russellian translation fogs our understanding of the sentence, which is given in the proof: 'The curse of the invasion of mathematics by mathematical logic is that now any proposition can be represented in a mathematical symbolism, and this makes us obliged to understand it. Although of course this method is nothing but the translation of vague ordinary prose' (*RFM* v, § 47). See also (*RFM* iii, § 25).

and brings in a system of propositions, for instance of propositions of the form $a \times b = c$. [. . .]

The method of checking its truth corresponds to the sense of a mathematical proposition. If it's impossible to speak of such a check, then the analogy between 'mathematical proposition' and the other things we call propositions collapses. (*PG*, p. 366)

By linking sense with possession of a 'method of checking' or decision procedure, Wittgenstein was making room for the possibility of an error in the application of the procedure. It was thus possible to recognize some 'sense' in propositions such as $25 \times 25 = 620$, contrary to what he had claimed, when he was still thinking (as we saw in section 4.1) in terms of the *Tractatus Logico-Philosophicus*: As pointed out then, if one cannot imagine under which circumstances one would be prepared to assert the negation $\neg F(a)$, one cannot attach a clear sense to '$F(a)$ might be false', and questions such as 'Can $F(a)$ be revised?' have no sense: 'Where you can't look for an answer, you can't ask either' (*PR*, § 149; *PG*, p. 377). Linking sense with the availability of a 'method of checking' allows Wittgenstein, however, to move away from this strong conclusion:

A doubt whether $25 \times 25 = 620$ (or whether it $= 625$) has no more no less sense than the method of checking gives it. It is quite correct that we don't here imagine, or describe, what it is like for 25×25 to be 620. (*PG*, p. 392)

Whether an expression has sense depends upon the calculus. I can imagine the kind of mistake which could lead one to say $26 \times 13 = 1560$, or that 4 is the first digit of π, and thus I could say that the corresponding questions about them are genuine. (*AWL*, p. 200)

Wittgenstein had in mind throughout elementary equations of the form $a \times b = c$ but a better example would perhaps be propositions such as 'every polynomial of degree 1 has a root' or 'every polynomial of degree 2 has a root', which are decidable (while the general proposition 'every polynomial has at least one root' is not).[11] The proposition 'every polynomial of degree 1 has a root' is part of a system of elementary algebra for which a procedure of decision, based on a generalization of Sturm's theorem, was discovered by Tarski (1951). So, according to Wittgenstein, the 'sense' of the proposition 'every polynomial of degree 1 has a root' is given by this procedure

[11] This example is actually somewhat inadequate, since the sentence 'every polynomial of degree 1 has a root' is already a general one. But this is of little importance here.

of decision. We can now see clearly that for Wittgenstein understanding of a mathematical proposition is based on knowledge or possession of *decision procedure* for a given system of propositions—here elementary algebra—within which propositions such as 'every polynomial of degree 1 has a root' can be meaningfully asserted. Decidability of the system of elementary algebra ensures the applicability of the Law of Excluded Middle as well as other laws of classical logic, and an obvious inference is that Wittgenstein believed that these laws are indeed applicable.[12] In that sense, to use to my own purpose Frascolla's words, a '*qualified* use of the word "proposition" is justified' (Frascolla 1994: p. 63).

One final remark: since a 'method of checking' or a decision procedure is nothing more than an algorithm, one is led once more to the fundamental (and very general) feature of Wittgenstein's philosophy of mathematics, namely the view of mathematics as being essentially algorithmic or as some sort of 'abacus' or 'calculating machine' that 'works by means of strokes, numerals' (*WVC*, p. 106), which was presented in Chapter 1.

These considerations lead us naturally to the topic of the validity of the Law of Excluded Middle, to which I shall turn now. To use the distinction between assertions and hypotheses or inductions introduced in section 5.2 above, it seems as if Wittgenstein is ready to admit decidability (and hence the applicability of the Law of Excluded Middle) at the level of assertions (mathematical propositions) when there is a procedure of decision available (therefore a 'calculus') but not at the level of the expression of the procedures of decision themselves. This is a very difficult issue, and I hope that the following remarks will help to settle a few points once and for all.

6.2. EXCLUDED MIDDLE AND EXISTENCE

Throughout his career, Brouwer defended the view not only that mathematics is essentially a languageless activity but that the mental constructions of the mathematician can only be imperfectly expressed in a linguistic form. This view, according to which mathematics is independent of logic, he set forth in details as early as 1907 in his doctoral thesis (Brouwer 1975: 72–97). A year later he published a short

[12] This is the conclusion reached by Frascolla (1984; 1994: 62–3).

paper on 'The Unreliability of Logical Principles' in which he asked which of the principles of classical (Aristotelian) logic remain reliable when extended to the domain of the 'mathematics of the infinite' (*Unendlichkeitsmathematik*): Can one start from the linguistic expression of a mathematical construction, apply the principles of logic, and still obtain a reliable result when returning to the mathematical construction? Brouwer's verdict was that reliable results will be obtained if the 'principle of syllogism' (inclusion of classes) and the 'principle of contradiction' are used, but not with the 'principle of *tertium exclusum*' or Law of Excluded Middle (p. 110). Twenty years later he was saying exactly the same in front of Wittgenstein in Vienna: 'it seems that there is in general no mathematical reality corresponding to assertions of [the Law of Excluded Middle] or to inferences based on it' (1975: 425). In the 1920s, Brouwer articulated his criticism of the universal validity of the Law of Excluded Middle around his now typical example of the decimal expansion of π and around his pendulum number (*duale Pendelzahl*).

Many commentators on the later Wittgenstein see him as a critic of the universal validity of the Law of Excluded Middle. For example, Robert Fogelin lists his 'attacks upon the unrestricted use of the Law of Excluded Middle' as one of the intuitionist themes found in Wittgenstein's writings (Fogelin 1968: 267). Commentators with anti-realist leanings are more careful: Michael Dummett speaks of the 'ambivalence' of Wittgenstein's attitude towards the Law of Excluded Middle.[13] Crispin Wright claims that no explicit rejection of it is to be found in Wittgenstein's writings, and that one might find instead some shared common background for its rejection (Wright 1980: 142). The very idea that Wittgenstein had any qualms about the universal validity of the Law of Excluded Middle is, however, strongly rejected by many commentators whose analyses are based on the analogy between ordinary and mathematical propositions. The view is that to be decidable defines what it is to be a proposition, so where the Law of Excluded Middle does not apply one simply cannot talk of a proposition. If mathematical equations are analogous in a strong sense to propositions, then the very idea of an undecidable

[13] According to Dummett (1978a: 178–9), Wittgenstein's attitude is ambivalent because on the one hand insistence on the Law of Excluded Middle is the expression of a Platonist outlook (according to which the general form of an explanation of meaning is in terms of truth-conditions) which he repudiated and, on the other hand, he wished not to interfere with the mathematician: if the latter wishes to use a form of argument based on the Law of Excluded Middle, he has the right to do so.

mathematical proposition is simply nonsense. There is support for such a view in Wittgenstein's remarks on Brouwer and Weyl: he nearly always puts the emphasis on differences between his views and theirs. Indeed, his reaction to their criticisms of the universal applicability of the Law of Excluded Middle appears at first sight to be negative: 'I need hardly say that where the law of the excluded middle doesn't apply, no other law of logic applies either, because in that case we aren't dealing with propositions of mathematics. (Against Weyl and Brouwer.)' (*PR*, § 151).[14] Such a reading, often motivated by the desire to avoid the conclusion that Wittgenstein adopted an unpopular stance on a controversial thesis, simply stops at the negative impression such remarks convey; it is not very insightful. I have already presented a reconstruction of Wittgenstein's position on general arithmetical propositions in section 4.2, which implied that he objected to at least the universal applicability of the Law of Excluded Middle; I shall now provide further arguments in support of my interpretation.

I should begin by pointing out a set of remarks that can readily be cited in support of the claim that Wittgenstein objected to at least the universal applicability of the Law of Excluded Middle, namely his discussion of Brouwer's example of the appearance of the pattern 123456789 in the decimal expansion of π. There is here much agreement between Wittgenstein and intuitionists. Brouwer's example is a formula of the type:

$$\exists x\ F(x) \lor \forall x \neg F(x)$$

with x ranging over an an infinite series. Intuitionists would say here that one needs to show either that one can effectively find an a such that $F(a)$ or that one cannot find one. There are claims one cannot make about certain decimal expansions, such as that of π, unless the calculation up to the relevant point is already done, because there is not enough information given with the rule with which one computes it. Therefore, unless one already finds, for example, the pattern 123456789 in one's computation of the decimal expansion of π, one cannot assert the corresponding $\exists x\ F(x)$ because this claim would be unjustified. Moreover, the fact that one does not possess an instance of $F(a)$ does not allow one to assert $\forall x \neg F(x)$. Wittgenstein clearly followed intuitionists on this score; such a reasoning is behind the remark recorded by Waismann in December 1929:

[14] See also *M*, p. 302.

> There can be no such question as, Do the figures 0, 1, 2 . . . 9 occur in π? I can only ask if they occur at *one* particular point, or if they occur among the first 10,000 figures. No expansion, however far it may go, can refute the statement 'They do occur'—therefore this statement cannot be verified either. What is verified is an entirely different assertion, namely that this sequence occurs at *this or that point*. Hence you cannot affirm or deny such a statement, and therefore you cannot apply the law of the excluded middle to it. (*WVC*, p. 71)

Since the issue is highly controversial, it is worth pointing out other passages where a similar claim was made. Wittgenstein is quoted by Alice Ambrose as saying, during the academic year 1931–2:

> Will three consecutive sevens ever occur in an evaluation of π? People have an idea that this is a problem because they think that if we knew the whole evaluation we should know, and the fact that we don't know is merely a human weakness. This is a subterfuge. The mistake lies in the misuse of the word infinite, which is not the name of a numeral.
>
> 'If we find that three consecutive sevens occur, then we have proved that they do; but if we don't find them we still have not proved that they do not.' This gives us no criterion for falsehood, but only for truth. (*AWL*, p. 107)

(It is worth noticing here the connection with Wittgenstein's views on infinity, to be discussed in the next section.) Moore also reported that Wittgenstein

> said that if anyone actually found three consecutive 7's this would prove that there are, but that if no one found them that wouldn't prove that there are not; that, therefore, it is something for the truth of which we have provided a test, but for the falsehood of which we have provided none; and that therefore it must be a quite different sort of thing from cases in which a test for both truth and falsehood is provided. (*M*, p. 303)

It seems clear that Wittgenstein had a good grasp of the intuitionists' argument and, at the risk of denying the above evidence, one must admit that he agreed with it.

There are very interesting remarks where Wittgenstein agrees with Brouwer's rejection but rejects his line of argumentation and offers his own instead. The first one involves Brouwer's pendulum number, about which I should say a few words. Wittgenstein had first-hand knowledge of it, since Brouwer presented it during the Viennese lecture which he attended. It is constructed thus: let d_v be the vth digit of the decimal expansion of π and $m = k_n$ if at dm for the nth time the segment $d_m d_{m+1} \ldots d_{m+9}$ forms the sequence 0123456789. Now, if

$v \geq k_1$ then $cv = (-1/2)^{k_1}$, otherwise $cv = (-1/2)^v$. The sequence c_1, c_2, c_3, . . . forms the real number r—the so-called pendulum number—for which it is impossible to tell if $r = 0$, $r > 0$ or $r < 0$ (Brouwer 1967a: 337). This number has some peculiar properties: in Brouwer's own words in Vienna, it is 'not rational, although its irrationality is absurd, and not comparable with zero, although its incomparability with zero is absurd' (Brouwer 1975: 163).[15] Brouwer's intention in building such a number was to provide a counterexample to the Law of Excluded Middle. He claimed in Vienna that 'this pendulum number is neither equal nor unequal to zero, in contradiction with the Law of Excluded Middle' (Brouwer 1975: 161). To this affirmation, Wittgenstein replied:

> Brouwer is right when he says that the properties of his pendulum number are incompatible with the law of the excluded middle. But, saying this doesn't reveal a peculiarity of propositions about infinite aggregates. Rather, it is based on the fact that logic supposes that it cannot be *a priori*—i.e. logically—impossible to tell whether a proposition is true or false. For, if the question of the truth or falsity of a proposition is *a priori* undecidable, the consequence is that the proposition loses its sense and the consequence of this is precisely that the propositions of logic lose their validity for it.
>
> Just as in general the whole approach that if a proposition is valid for one region of mathematics it need not necessarily be valid for a second region as well, is quite out of place in mathematics, completely contrary to its essence. Although these authors hold just this approach to be particularly subtle, and to combat prejudice. (*PR*, § 173)

Wittgenstein's reasoning seems to be this: if the applicability of logic requires that it is a priori possible to tell if the proposition is true or false, then where that requisite is not fulfilled, logic does not apply. One may choose to read Wittgenstein as saying that the idea of an undecidable proposition is thus nonsense and that Brouwer's conception must be incorrect: if there are mathematical propositions, and there are, then logic must apply. It is, however, this reading which is plainly incorrect. It is clear that Wittgenstein holds Brouwer to be right, but for the wrong reason. The above passage says no more than this: Brouwer is right that the Law of Excluded Middle does not apply, but errs in believing that it is the case because of a peculiarity

[15] It is precisely on that account that Wittgenstein rejected Brouwer's pendulum number as not being a true real number (*WVC*, p. 73). Indeed, Wittgenstein made it an essential requirement that every real number be effectively comparable with any rational number. I shall come back to this question in section 7.1.

of infinite sets that he has uncovered. Wittgenstein then supplies his own argument, which is (to repeat) that the applicability of logic requires that it is a priori possible to tell if the proposition is true or false and that otherwise logic does not apply.

It seems to me that Wittgenstein rejected Brouwer's argument against the universal applicability of the Law of Excluded Middle because he sensed in Brouwer's discourse a remnant of Platonism. According to him, Brouwer spoke as if he had just discovered some special fact about a certain class of propositions (those about infinite sets) in the same way as a physicist would speak of discovering the laws of nature. I shall cite this long but instructive passage of the *Philosophical Grammar* (but see also *AWL*, p. 140) where, again, Wittgenstein clearly agreed with Brouwer's stance on the Law of Excluded Middle but not with the way he defends it:

> When Brouwer attacks the application of the law of excluded middle in mathematics, he is right in so far as he is directing his attack against a process analogous to the proof of empirical propositions. In mathematics you can never prove something like *this*: I saw two apples lying on the table, and now there is only *one* there, so A has eaten an apple. That is, you can't by excluding certain possibilities prove a new one which isn't already contained in the exclusion because of the rules we have laid down. To that extent there are no genuine alternatives in mathematics. If mathematics was the investigation of empirically given aggregates, one could use the exclusion of a part to describe what was not excluded and in that case the non-excluded part would not be equivalent to the exclusion of the others. The whole approach that if a proposition is valid for one region of mathematics it need not necessarily be valid for a second region as well, is quite out of place in mathematics, is completely contrary to its essence. Although many authors hold just this approach to be particularly subtle and to combat prejudice. (*PG*, p. 458)

One can indeed read remarks such as these as containing an argument against what Wittgenstein saw as the reintroduction by the back door of a Platonist way of thinking in Brouwer's arguments against the Law of Excluded Middle. Thus Wittgenstein can be seen merely to be arguing against Brouwer's sales pitch. It appears to me that Wittgenstein clearly misconstrued Brouwer's critique by describing it as implying the Platonist viewpoint, since the point of Brouwer's counterexamples, such as the pendulum number, is precisely to avoid that viewpoint. Brouwer's ultimate intention was to show that incompletely defined mathematical objects, such as choice sequences, require another underlying logic (see section 7.2). The pendulum number is a perfect counterexample, since it is given by a constructive function—the trick is that it is given in such way that nobody can determine from the rule only whether or not the number r is equal to 0. There is no underlying Platonism here.

More importantly, Wittgenstein's point is not just against Brouwer's alleged Platonism. His claim is that the *real* reason why the Law of Excluded Middle is not valid is, again, that the applicability of logic requires that it is possible to tell a priori whether the proposition is true or false, and that otherwise logic does not apply—a claim which he expressed in the above quotation by saying that there are 'no genuine alternatives in mathematics'. Wittgenstein's stance is much more radical than Brouwer's—this is an issue that I skirted in section 4.2 above—for he claims in both passages (*PR*, § 173; *PG*, p. 458) that the lack of validity of the Law of Excluded Middle in mathematics is a distinguishing feature of *all* mathematical propositions (as opposed to empirical propositions) and not only a particularity of the mathematics of the infinite. If anything, Wittgenstein was all along not so much doubting the reason set forth by Brouwer against the universal applicability of the Law of Excluded Middle as arguing for its universal *in*applicability.

This radical stance is rooted in the *Tractatus Logico-Philosophicus*, where the validity of the Law of Excluded Middle appears, at first sight, to be linked with his belief in the 'bipolarity' of propositions (there is no bipolarity if it is impossible to negate the proposition—*NB*, p. 94). According to the *Tractatus Logico-Philosophicus* mathematical equations lack 'bipolarity', and this is in turn why Wittgenstein described them as *Scheinsätze* (*TLP*, 6.2). We saw in section 4.2 above that this much remained unchanged, at least initially, during the transitional period: Wittgenstein was still holding that mathematical propositions cannot be negated in the ordinary sense of the word. According to him, 'it is part of the nature of *what we call propositions* that they must be capable of being negated' (*PG*, p. 376). So the negation of a meaningful proposition must also be meaningful, and we saw in section 4.2 above that Wittgenstein thought that mathematical equations cannot be negated in the ordinary sense of the word because one cannot imagine under what circumstances we would be prepared to assert the negation $\neg F(a)$ of a free-variable formula F, once we proved it. He thus wrote that '[i]t is quite clear that negation in arithmetic is completely different from the genuine negation of a statement' (*PR*, § 202) and that '[n]egation in arithmetic cannot be the same as the negation of a statement' (*PR*, § 203).

In section 4.2, I also pointed out the parallel with Hilbert's conception according to which the sentential operators had to be truth-functional, so that they could intelligibly apply only to propositions with determinate truth-conditions. Precisely because justification-conditions and not truth-conditions can be associated with general propositions involving unrestricted quantification, Hilbert admitted that 'those laws that man has always used since he began to think, the very ones that Aristotle taught, do not hold' (Hilbert 1967b: 379). This is an important consequence of the finitist account: it is not just the Law of Excluded Middle that does not apply to finitary propositions but the whole calculus of truth-functions. Wittgenstein's conception is in the same vein but more radical. He argued too that his calculus of truth-functions in the *Tractatus Logico-Philosophicus* does not apply to mathematical equations. This is in line with the finitist account about general propositions involving unrestricted quantification, but it is also more radical because he extended the claim to elementary, 'real' arithmetical propositions.

Intuitionistic logic had not yet been formalized in the 1920s, when Wittgenstein published his *Tractatus Logico-Philosophicus*; and one had to wait for Heyting's work mentioned at the beginning of the last section. Following it, intuitionists, while accepting the finitist account of quantifiers, believe that one can explain the meaning of the sentential operators not by truth-tables but by specifying for each one of them what would justify an assertion of a statement of which it is the principal operator. We should view Wittgenstein as clearly siding against intuitionists: since he was at the time thinking of the logical connectors in terms of truth-tables, as in the *Tractatus Logico-Philosophicus*, the fact that the Law of Excluded Middle does not hold implied, for him, that 'no other law of logic applies either'. This is how one should read § 151 of the *Philosophical Remarks* already quoted: 'I need hardly say that where the law of the excluded middle doesn't apply, no other law of logic applies either, because in that case we aren't dealing with propositions of mathematics. (Against Weyl and Brouwer.)' (*PR*, § 151)

One more point should be made regarding § 173 of the *Philosophical Remarks*. One ought not to be puzzled by Wittgenstein's claim that 'the proposition loses its sense' if the Law of Excluded Middle does not apply. This is to be explained by the fact that in the *Tractatus Logico-Philosophicus* the conditions for 'sense' are intimately connected with possession of empirical content: 'I say: every possible proposition is legitimately constructed, and if it has no sense [*Sinn*] this can only be because we have given no *meaning* [*Bedeutung*] to some of its constituent parts' (*TLP*, 5.4733). Therefore, what lacks empirical content (*Bedeutung*) lacks sense (*Sinn*). What lacks empirical content also has no fact (no truth-maker) corresponding to it or to its negation; it is not bipolar, and hence not a proposition (*Satz*). (This is why not just mathematical equations but also propositions of metaphysics, ethics, and aesthetics are said to be lacking 'sense'.) So it would be, once more, diametrically opposed to the truth to see Wittgenstein's claim that 'the proposition loses its sense' if the Law of Excluded Middle does not apply as a defence of the latter against Brouwer's criticisms.

Wittgenstein's position could easily be contrued as being incoherent. There is indeed the appearance of contradiction between the above arguments for the inapplicability of the Law of Excluded Middle to all mathematical propositions and the conclusion, reached at the end of the previous section, that the Law of Excluded Middle is valid in decidable domains. I believe that there is only an appearance of a contradiction: as I have argued, the arguments for the inapplicability of the Law of Excluded Middle to all of mathematics are rooted in the conception of mathematical equations as *Scheinsätze* in the *Tractatus-Logico-Philosophicus*. In the early 1930s, Wittgenstein was simply in the process of changing his mind on this issue, introducing some legitimacy to the Law of Excluded Middle in contexts where a decision procedure is available.

Wittgenstein's remarks on mathematical existence are similar to those on the Law of Excluded Middle inasmuch as he also criticizes intuitionists' mode of speech while he in fact adopted their stance towards existence proofs. I shall turn to these briefly, but I would like first to express the hope of having clarified the general gist of Wittgenstein's position on the Law of Excluded Middle during the transitional period. There are many passages about the Law of Excluded Middle in later works such as the *Remarks on the Foundations of Mathematics* that I shall not discuss here. I would like to believe, however, that there is no significant departure from this line of thought. It has to be said, however, that the later Wittgenstein stops short, as Crispin Wright pointed out, of an outright rejection. But of the Law of Excluded Middle he said that it has a 'shaky' sense

(*RFM* v, § 12), that there was 'something wrong' in insisting on it (§ 10), and that it gives us a 'misleading picture' (§ 11; *PI*, § 352). These are more harsh words from someone who supposedly never doubted its validity.

Wittgenstein's discussion of mathematical existence is intimately linked with his view that the meaning of the theorem is to be found in the proof. I shall therefore avoid discussing mathematical existence in Hilbertian terms, i.e. in terms of consistency, but shall rather concentrate on the remarks on existence proofs. At the outset it is worth distinguishing, informally, between several kinds of proof. First one must distinguish between pure existence and constructive proofs. A proof of pure existence simply asserts the existence of a certain object, via a quantified statement $\exists x\, F(x)$, without giving the means—an algorithm—to construct or approximate it. Mere existence may be thought of as easier to establish than constructive existence but this does not necessarily mean that the result is easy to obtain.[16] It is often the case, however, that non-constructive proofs simply hide an algorithm and that they are constructivizable. For example, Errett Bishop has constructivized large parts of classical analysis in his book *Foundations of Constructive Analysis* (1967).[17]

Constructivity, however, comes in degrees. As Michael Beeson puts it, '[c]onstructivity is an attribute of proofs . . . that can be possessed in greater or lesser degree—it is not an all-or-nothing quantity like correctness' (1993: 139). To begin with, one ought to distinguish between proofs providing an algorithm for which no bound is known

[16] A recent example of a difficult existence proof is Gerd Faltings' proof of the Mordell–Weil conjecture (Faltings 1983: 365). Fermat's last theorem, which says that $X^n + Y^n = Z^n$, has no non-zero integer solution for $n \geq 3$, and Faltings' proof implies that for any $n \geq 3$ there can be at most a finite number of (pairwise co-prime) solutions. The proof provides no explicit bound z on the terms of n; we only know that *there is* such a bound.

[17] If Bishop's task were simply to reformulate known results in order to satisfy a philosophical craving for constructivity, it would have been, mathematically speaking, an unrewarding task. But Bishop's work brings improvements. In any case the need for more constructive proofs is a feature of mathematical practice, since the ultimate goal of mathematicians is complete solutions. Moreover, the lack of information in existential proofs is often a liability. The literature on Diophantine approximations based on the Thue–Siegel–Roth method, in transcendental number theory, is a perfect example, since it is the lack of constructivity of Thue's original existence result which plagued the field. Satisfactory results were obtained by Baker using another method. For a brief presentation, see Davenport (1982: 166–7), and for some details see Baker (1975: ch. 7).

from proofs providing bounded algorithms. The latter class of proofs is the one that satisfied most constructivists, such as Brouwer or Bishop. With the advent of the computer and the fast-growing literature on complexity theory, efficiency in the computation of algorithms has become a major preoccupation. As a result, an even more radical form of constructivism is by now taking shape (it is largely ignored by philosophers, although it is rapidly becoming the most popular approach): it is not enough that results be constructive, the algorithms have to be efficient enough to be run on computers, i.e. they need to be polynomially bounded. Thus, one ought to distinguish also between proofs providing polynomially bounded algorithms and other proofs, such as those with exponentially bounded algorithms. (In this section, this further distinction will play no role, but I shall allude to it briefly in section 8.2.) Equipped with this rough distinction between pure existence proofs and proofs with various degrees of constructivity, we can now look at Wittgenstein's remarks. Recall that for Wittgenstein there is a danger in focusing on the 'prose' or its translation in Russellian notation, instead of looking at the proof, i.e. at the calculations (see *PG*, pp. 369–70, quoted above, and *PR*: § 163). One needs more to be able to understand a mathematical proposition than a mere 'verbal' understanding. I shall now jump for a moment to later remarks, taken mainly from part v of the *Remarks on the Foundations of Mathematics*, written around 1942–4. Wittgenstein had by then introduced his notion of the 'language-game', in terms of which he now couches his point: 'One would like to say that the understanding of a mathematical proposition is not guaranteed by its verbal form, as in the case with most non-mathematical propositions. This means—so it appears—that the words don't determine the language-game in which the proposition functions' (*RFM* v, § 25). This remark is linked to Wittgenstein's general distrust of the intrusion of mathematical logic into mathematics. For him '[t]he Russellian signs veil the important forms of proof as it were to the point of unrecognizability, as when a human form is wrapped up in a lot of cloth' (*RFM* iii, § 25); the logical notation 'swallows the structure' (v, § 25).[18] All this is reminiscent of Brouwer's distrust of the language of mathematics. But the reasons

[18] Wittgenstein also wrote: 'The harmful thing about logical technique is that it makes us forget the special mathematical techniques. Whereas logical technique is only an auxiliary technique in mathematics. . . . It is almost as if one tried to say that cabinet-making consisted in glueing' (*RFM* v, § 24).

for not trusting the verbal form are different: Brouwer would claim in a very solipsistic manner that language can only render imperfectly the mathematician's thoughts, while for Wittgenstein the essential aspect of a proof is the recipe or algorithms that it provides, and there is a danger in logic of seeing this aspect as inessential:

> When a proof proves in a general way that *there is* a root, then everything depends on the form in which it proves this. On what it is that here leads to this verbal expression, which is a mere shadow, and keeps mum about *essentials*. Whereas to logicians it seems to keep mum only about incidentals. (*RFM* v, § 25)

It is precisely in this context that Wittgenstein questioned the value of pure existential proofs. Here are two telling quotations:

> Hence the issue whether an existence-proof which is not a construction is a real proof of existence. That is, the question arises: Do I understand the proposition 'There is . . .' when I have no possibility of finding where it exists? And here are two points of view: as an English sentence for example, I understand it, so far, that is, as I can explain it (and note how far my explanation goes). But what can I do with it? Well, not what I can do with a constructive proof. And insofar as what I can do with the proposition is the criterion of understanding it, thus far it is not clear in advance whether and to what extent I understand it. (*RFM* v, § 46)

> A proof that shews that the pattern '777' occurs in the expansion of π, but does not shew *where*. Well, proved in this way this 'existential proposition' would, for certain purposes, not be *a rule*. But might it not serve e.g. as a means of classifying expansion rules? It would perhaps be proved in an analogous way that '777' does not occur in π^2 but it does occur in $\pi \times e$ etc. The question would simply be: is it reasonable to say of the proof concerned: it proves the existence of '777' in this expansion? This can be simply misleading. It is in fact the curse of prose, and particularly of Russell's prose, in mathematics. (vii, § 41)

But these remarks date from the mid-1940s. During the transitional period, Wittgenstein insisted instead on the differences between his reasons for rejecting existence proofs and those of the intuitionists. For Wittgenstein, proof not only is the truth-maker, it is, so to speak, the meaning-maker. The very notion of mathematical existence is linked with that of proof: 'We have no concept of existence independent of our concept of an existence proof' (*PG*, p. 374). He simply pushed to the limit the intuitionist truth-maker analysis in terms of proofs and this led him, quasi-paradoxically, to reject the views about the existence of mathematical objects held by intuitionists.

According to Wittgenstein's radical stance, one cannot abstract from particular existential proofs a concept of 'existence'. As he would say, in a passage to be quoted in full below: 'What is an existential theorem? The answer is this, and this, and this . . .' (*AWL*, p. 116). According to this premiss it is impossible to obtain a unique concept of 'existence' after (and a fortiori to obtain one before) inspection of proofs with the help of which one could provide prescriptions. Wittgenstein believed that this was exactly what the intuitionists were, mistakenly, doing:

> 'Every existence proof must contain a construction of what it proves the existence of.' You can only say 'I won't call anything an "existence proof" unless it contains such a construction'. The mistake lies in pretending to possess a clear *concept* of existence.
>
> We think we can prove a something, existence, in such a way that we are then convinced of it independently of the proof. (The idea of proofs independent of each other—and so presumably independent of what is proved.) Really, existence is what is proved by the procedures we call 'existence proofs'. When the intuitionists and others talk about this they say: 'This state of affairs, existence, can be proved only thus and thus.' And they don't see that by saying that they have simply defined what *they* call existence. For it isn't at all like saying 'that a man is in the room can only be proved by looking inside, not by listening at the door'.
>
> We have no concept of existence independent of our concept of an existence proof. (*PG*, p. 374)

This objection was made again during the lectures of Michaelmas Term 1934:

> Weyl said that every existential proof must consist in constructing what is said to exist. But *must* it? Doesn't this depend on what is called an existential proof? Weyl is using the fact that in a huge number of cases something is done which might be called constructing a certain entity. What is an existential theorem? The answer is this, and this, and this . . . If there were such a thing as existence which is proved when an existence theorem is proved, then perhaps one could say every existential proof must do a certain thing. Weyl talks as though he has a clear idea of existence independent of proof, and has made what looks as a statement about the natural history of proofs in saying that only such-and-such prove existence. There is no concept of existential theorems except through the special existential theorems. Every existential proof is different, and 'existential theorem' has different meanings according as what is said to exist is, or is not, constructed. Of course one can arbitrarily fix a criterion: one can call an existential proof one which fulfills certain formal conditions. (*AWL*, pp. 116–17)

Wittgenstein shared broadly the intuitionist lack of satisfaction with existence proofs. What is left for him to criticize is the language in which they couched their remarks. For him there is no concept of existence independent of particular proofs, and 'Weyl talks as though he has a clear idea of existence independent of proof'. It looks, according to Wittgenstein, as if Weyl and Brouwer are making statements 'about the natural history of proofs'—something he strongly disagrees with: 'Confusion in these matters are entirely the result of treating mathematics as a kind of natural science' (*PG*, p. 375). So the presumed intuitionist prescription 'Every existence proof must contain a construction of what it proves the existence of' must be replaced by the more appropriate statement 'I won't call anything an "existence proof" unless it contains such a construction.'

Wittgenstein's strong words against the language of Brouwer and Weyl are likely to confuse some into thinking that his position is radically opposed to that of the intuitionists. But this is not the case: his criticisms of their language in no way implies the rejection of their critical standpoint. If it did, one would have to explain away, on pain of contradiction, Wittgenstein's later remarks on existence, with their clear constructivist slant and the underlying distrust of Russellian mathematical logic. A more interesting point to emphasize is the similarity between Wittgenstein's criticisms of the language in which intuitionists couch both their criticisms of the universal validity of the Law of Excluded Middle and their prescriptions about mathematical existence. The intuitionists seemingly base their arguments on a prior stance on the existence of mathematical objects, which differs from that of the Platonists. Wittgenstein disagreed with both intuitionists and Platonists on that score. We shall see in the next section that from his 'intensional' standpoint, intuitionists and Platonists are indeed alike in that they assume the existence of extensions and argue, so to speak, about their size.

6.3. FORMALISM, INFINITY, AND EPISTEMIC LIMITATIONS

The third part of Frege's *Grundgesetze der Arithmetik* contains a lengthy critical discussion of the foundational work of J. Thomae and E. Heine (Frege 1980b: §§ 86–137), in which Frege puts forth an impressive array of arguments (on average one per section). In some passages, Frege needlessly picks on vague formulations, in others he uses powerful arguments which appeal to central elements of his doctrine, such as the argument about the applicability of arithmetic (§ 91). Frege also shows in these sections that he had a grasp of the use/mention distinction (§ 98). The issue is for him between 'formal' arithmetic (hence the name 'formalism' associated with those defending it) and arithmetic 'with content' (*inhaltliche*) (§ 88). Formal arithmetic, which is set forth in the writings of Thomae and Heine under consideration, is the view that arithmetic is a manipulation of meaningless signs, while arithmetic with content, Frege's own standpoint, is the view according to which equations are sentences expressing thoughts (Frege's *Gedanken*). So the issue for Frege is this: either we have 'dead' signs (as Wittgenstein called them in *BB*, p. 4) or we recognize that the signs denote an independent and pre-existing reality.

One peculiarity of the formalism of Thomae and Heine is their insistence on the tangibility of the signs.[19] These are simply marks of ink on paper or of chalk on a blackboard. Therefore, blackboard and chalk are no longer external aids to the mathematician but 'essential constituents of the theory itself' (Frege 1980b: § 87). This move was conceived as a way of empirically (not logically) ascertaining the existence of numbers. Indeed, if numbers are tangible signs, then their existence is as certain as the signs are tangible. For Frege, this idea is deeply confused. One area where the confusion becomes patent is infinity. Infinite sequences of natural numbers are obviously needed (for the introduction of irrationals, and so forth), and these are defined as sequences having no last member, so that according to the their rule of construction, new terms can always be constructed. But, as Frege ironically pointed out, this is simply impossible if one holds the view that numbers are tangible signs:

We gather from Heine's exposition that such series is to continue to infinity. In order to produce it we would need an infinitely long blackboard, an infinite supply of chalk, and infinite length of time. We may be censured as too cruel for trying to crush so high a flight of the spirit by such a homely objection; but this is no answer. If numbers are taken to be tangible figures, whose existence is rendered certain by their tangibility, why then they must be subject to

[19] There are differences between Thomae and Heine that are overlooked here. For example, Heine confused number and numeral. The obvious objection here is that written signs have properties that we would not wish to ascribe to numbers. Thomae avoided this objection by claiming, in the quotation below, that arithmetic is not a theory of the written signs but a game played with them.

all the limitations of such material existence. We see that Heine is the victim of a curious fate: the tangibility of numbers, which is supposed to guarantee the existence of numerical series, and consequently of irrational numbers, is in fact just what makes their existence impossible. (Frege 1980b: § 124)

Indeed, Frege argued, the formalist view implies on the one hand that an infinite series is in some sense concretely written down (otherwise there would not be such a thing) and on the other that, because one humanly cannot write down an infinite series, there would be no last term to that series; it could not be concretely written down. The series would then remain as a possibility for God to realize:

The emphasis in Thomae's definition . . . is upon possibility . . . in accordance with a given prescription new terms and more new terms *can* always be constructed. But new terms and more new terms need not actually be constructed. The possibility is enough . . . Does the possibility exist? For an almighty God, yes; for a human being, no. (Frege 1980b: §125)

There is indeed no more sense in talking about an infinite row of houses than in talking about an infinite series of numbers conceived in the manner of the formalists: we cannot admit of a row of houses that it is endless, neither can we admit of such a series of concretely written numbers that it is infinite. According to Frege, the formalist conception runs into absurdities and this proves, by *reductio ad absurdum*, the rightness of his idea of an 'arithmetic with content'. These difficulties disappear when one adopts the view that in mathematics signs stand for a reality: one can then refer to the infinite with finite signs, provided the proper definitions are adopted. For example, the number of the set of all finite cardinals—Frege's equivalent of Cantor's $\aleph_0$—has an identical status to that of any finite cardinal, since it is defined according to the same model (Frege 1980a: § 84).

In obvious reference to Frege's *Grundgesetze der Arithmetik*, Wittgenstein pointed out, in conversation with Schlick and Waismann, that there is a third possibility, alongside the views of the formalists and Frege:

For Frege the alternative was this: either we deal with strokes of ink on the paper or these strokes of ink are signs *of something* and their meaning is what they go proxy for. The game of chess itself shows that these alternatives are wrongly conceived—although it is not the wooden chessmen we are dealing with, these figures don't go proxy for anything, they have no meaning in Frege's sense. There is still a third possibility, the signs can be used the way they are in a game. (*WVC*, p. 105)

The accusation is certainly unfair, since Frege was aware of the analogy with the game of chess; it was made by Thomae in a passage which Frege himself quotes and discusses at length (1980b: §§ 88 f.). This passage is worth putting side by side with Wittgenstein's remark:

> For the formalist, arithmetic is a game with signs, which are called empty. That means they have no other content (in the calculating game) than they are assigned by their behaviour with respect to certain rules of combination (rules of the game). The chess player makes similar use of his pieces; he assigns them certain properties determining their behaviour in the game, and the pieces are only external signs of this behaviour. (Frege 1980b: § 88)

According to Thomae, signs in arithmetic derive their meaning from the rules of arithmetic. This sounds strikingly similar to Wittgenstein's position as stated above, but there could not be an idea more alien to Frege than this. For Frege the situation ought to be exactly the reverse: rules are derived from meanings. As he construed the formalist position, rules are simply created, stipulated arbitrarily (Frege 1980b: § 89); they have no basis: 'We do not derive these rules from the meaning of the signs, but lay them down on our own authority, retaining full freedom and acknowledging no necessity to justify the rules' (§ 94). Frege even accuses Thomae of presupposing meanings in his formal arithmetic:

> Although numerical signs designate something, this can be ignored, according to Thomae, and we can regard them simply as pieces manipulated in accordance with rules. If their meaning were to be considered, this would supply the grounds for the rules; but this occurs behind the scenes, so to speak, for on the stage of formal arithmetic nothing of the sort can be seen. (§ 90)

Wittgenstein was not in any strict sense a formalist. For example, one aspect of the formalist doctrine which has no equivalent in his writings is the insistence on the tangibility of signs: 'Is Mathematics about signs on paper? No more than chess is about wooden pieces' (*PG*, p. 289). But it is no exaggeration to say that the whole of Wittgenstein's later philosophy of language goes against the *Bedeutungskörper* conception of meaning upon which Frege's 'arithmetic with content' is based.[20] His approach is antithetical to Frege's:

[20] In fact, the first part of the *Philosophical Investigations* (*PI*, §§ 1–138) is but a lengthy critique of this conception. For an excellent account, see Bouveresse (1987: 30–9).

'Rules do not follow from an act of comprehension' (*AWL*, p. 50). When speaking about comprehension, Frege's usual turn of phrase is that we 'grasp' (*Fassen*, *Erfassen*) meanings, but he never explained what he meant by that. In *Philosophical Grammar*, Wittgenstein pointed out this lacuna:

> In attacking the formalist conception of arithmetic, Frege says more or less this: these petty explanations of the signs are idle once we *understand* the signs. Understanding would be something like seeing a picture from which all the rules followed, or a picture that makes them all clear. But Frege does not seem to see that such a picture would itself be another sign, or a calculus to explain the written one to us. (*PG*, p. 40)

The following passage presents Frege's conception, as Wittgenstein understood it: 'It looks as if one could *infer* from the meaning of negation that "$\neg\neg$ p" means p. As if the rules for the negation sign follow from the nature of negation. So that in a certain sense there is first of all negation, and then the rules of grammar' (*PG*, p. 53). Wittgenstein wanted to reverse Frege's conception and he argued that rules do not follow from but create, so to speak, meanings. His view was set forth in a conversation with Schlick and Waismann:

> when we are talking about negation, for instance, the point is to give the rule '$\neg\neg$ p = p'. I do not assert anything. I only say that the structure of the grammar of '$\neg$' is such that 'p' may be substituted for '$\neg\neg$ p'. Were you not also using the word 'not' in that way? If that is admitted, then everything is settled. And this is how it is with grammar in general. (*WVC*, p. 184)

For Wittgenstein, there is a difference between asserting the rule $\neg\neg p = p$ as if it were a discovery of some fact about the meaning of negation (this would be a typical example of what he called the metaphysician's confusion between a grammatical inquiry and an empirical one (*Z*, § 458)) and simply giving it as a grammatical rule.[21] Such rules are 'all the conditions necessary for the understanding (of the sense)' (*PG*, p. 88). It is fitting to notice that in this passage the connection with the newly found ideas on rules of grammar (the heir of the hypotheses of 1929, as explained in section 5.2 above) is clearly stated: a rule of grammar is not a statement; it does not assert anything.

[21] Readers should be aware of the relation here with Wittgenstein's non-cognitivist view of philosophy, i.e. the view that in philosophy one does not search for new facts (*WVC*, p. 183; *PI*, § 89). See Marion (1993).

To come back to the topic of infinity, while Wittgenstein did not deny the incoherence of the formalist view, he had to show that we need not adopt Frege's 'arithmetic with content'. According to Wittgenstein, both Frege and the formalists make the same mistake. They both admit that with our finite means, we cannot grasp the infinite. The only difference is that the formalist needs God to write down what she, as a human, cannot write down; while Frege claims that there exists an infinite and one can refer to it with finite means. For Wittgenstein, Frege's approach was plainly wrong: 'Set theory attempts to grasp the infinite at a more general level than the investigation of the laws of the real numbers. It says that you can't grasp the actual infinite by means of mathematical symbolism at all and therefore it can only be described and not represented' (*PG*, p. 468). Since Wittgenstein thought that 'we can't describe mathematics, we can only do it' (*PR*, § 159), to conceive a calculus as a description was according to him a complete misunderstanding of its nature, as we saw in Chapter 1.

Coupled with the view that mathematical truths are timeless—'numbers proper are eternal' (Frege 1980b: § 131)—Frege's approach is tantamount to seeing the infinite in mathematics as 'actual'. One ought not to infer, however, from the contradictions of the formalist viewpoint that the infinite has nothing to do with the 'possible', that it is not 'potential' but in a certain sense 'actual'. There is, according to Wittgenstein, a danger in giving the status of 'reality' to the 'possible': 'The word "possibility" is of course misleading, since someone will say, let what is possible now become actual. And in thinking this, we always think of a temporal process and infer from the fact that mathematics has nothing to do with time, that in its case possibility is (already) actuality' (*PR*, §141; *PG*, p. 471). Wittgenstein thought that it is 'one of the most deep rooted mistakes of philosophy to see possibility as a shadow of reality' (*PG*, p. 283).[22] One must try, therefore, to avoid describing an infinite series such as the natural numbers as possessing a 'kind of shadowy reality':

> Of course the natural numbers have only been written down up to a certain highest point, let's say 10^{10}. Now what constitutes the *possibility* of writing down numbers that have not yet been written down? How odd is this feeling that they are all somewhere already in existence! (Frege said that before it was drawn a construction line was in a certain sense already there.)

[22] There is, of course, the opposite 'danger of falling into a positivism' (*PG*, p. 283), which would consist of admitting, as the formalists do, only what is concretely given.

The difficulty here is to fight off the thought that possibility is a kind of shadowy reality. (*PG*, p. 281) [23]

Wittgenstein wanted to distinguish between an 'empirical' or 'physical' notion of possibility and a 'grammatical' or 'logical' notion, possibility;[24] the mathematical infinite being, according to him, of the latter kind. It is worth noticing that a similar distinction is to be found in Aristotle's discussion of the infinite: 'We should not take "being potentially" here as analogous to "this material is potentially a statue", which implies that it will in the future be a statue, and so conclude that there will in the future also be an actual infinite' (*Physics*, III, vi, 206a18–21). In this passage Aristotle is warning us not to confuse the kind of potentiality associated with the infinite with an empirical potentiality (as in the case of the statue). This is how one should understand the 'grammatical' infinite according to Wittgenstein: an expression ending with the words 'and so on' does not point towards a possibility waiting to be realized (an empirical possibility) but shows a possibility of the symbolism, i.e. it does not forbid us to continue: 'To say that a technique is unlimited does *not* mean that it goes on without ever stopping—that it increases immeasurably; but that it lacks the institution of the end, that it is not finished off' (*RFM* ii, § 45).[25] This 'possibility in the symbolism' (*LWL*, p. 17) is the indication, readable from the symbolism itself (*am Symbol allein*), that one must continue. Wittgenstein expressed this in typical fashion by likening the symbol $|0, \xi, \xi + 1|$ (see section 4.2 above) to an arrow, 'with the "0" as its tail and the "$\xi + 1$" as its tip: It is possible to speak of things which lie in direction of the arrow, but misleading or absurd to speak of all possible positions for things lying in the direction of the arrow as an equivalent for the arrow itself' (*PG*, p. 467).

Wittgenstein is here not very far from his earlier distinction between 'saying' and 'showing'. A rule does not *assert* that it is infinite; it is not a statement about an infinite extension. Rather, it *shows* it by allowing us to 'see an infinite possibility'.[26] So an infinite pos-

[23] In this passage Wittgenstein is referring again to the passage of the *Grundgesetze der Arithmetik* quoted in section 3.3 on numeral equivalence (Frege 1966: § 66).

[24] See e.g. *PG*, 452.

[25] See also *PR*, § 123; *PG*, 281–2; *RFM* v, § 14.

[26] We saw in section 4.2 that this distinction plays a significant role in Wittgenstein's account of general arithmetical propositions. It also helps to explain Wittgenstein's remarks on the continuum, in particular on Cauchy sequences, as will be shown in section 7.1.

sibility is expressed only in a rule or law of construction, not by a statement asserting it: 'Infinity is the property of a law, not of its extension' (*LWL*, p. 13). [27] This crucial point was made again in the following passage, taken from a short manuscript written in 1931: 'If we wish to say infinity is a attribute of possibility, not of reality, or: the word "infinite" always goes with the word "possible", and the like—then this amounts to saying: the word "infinite" is always part of a *rule*' (*PR*, app. i, p. 313). In other words, 'infinity' is a property of the rule for generating the natural numbers, not of the (completed) set of all natural numbers. It is only the possibility inscribed in the rule of continuing indefinitely to apply it; the rule does not forbid us to stop.[28] This conception is close to Aristotle's notion of *Apeiron*. For Aristotle, what is complete has an end (*telos*) and the end is a limit (*peras*). The infinite is incomplete, it has no end, no limit (as its name indicates: *Apeiron*, 'absence of limit'). The infinite may be regarded as the open possibility of more: 'Generally speaking, the infinite exists by one thing being taken after another. What is taken is always finite, but always succeeded by another part which is different from it' (*Physics*, III, vi, 206a27–8). For Aristotle, as for Wittgenstein, the infinite is what is lacking the 'institution of an end'.[29]

If the role played by the expression 'and so on' is to show the possibility of the symbolism, it has, so to speak, a completely different, 'strict and exact' grammar (*LWL*, p. 89), and it becomes of the

[27] See also *PR*, § 144 and app. 1, p. 313. At the same time, in Vienna, Felix Kaufmann expressed a similar finitist view: 'When we operate with the expression "and so on", as we do in describing infinite constructions, the stress must therefore lie on the word "so"; and that is determined by a law' (1930: 136). (Wittgenstein disagreed, it seems, with some aspect of Kaufmann's work, but Waismann's notes give us no clues: *WVC*, p. 84.)

[28] To speak of infinity as the property of an (infinite) extension is a confusion, which was brought about, Wittgenstein tells us, by a careless use of the word 'series'. He described the possibility of getting confused using the example of an infinite helix: 'Our normal mode of expression carries the seeds of confusion right into its foundations, because it uses the word "series" both in the sense of "extension", and in the sense of "law". The relationship of the two can be illustrated by a machine for making coiled springs, in which a wire is pushed through a *helically* shaped passage to make as many coils as desired. What is called an infinite helix need not be anything like a finite piece of wire, or something that that approaches the longer it becomes; it is the law of the helix, as it is embodied in the short passage. Hence the expression "infinite helix" or "infinite series" is misleading' (*PG*, p. 430).

[29] The point of the comparison with Aristotle is simply to show that Wittgenstein's 'grammatical' conception rests squarely within the Aristotelian tradition, as opposed to Cantor's.

utmost importance that it should not be understood as an abbreviation for an infinite extension (the three dots must not be understood as 'dots of laziness', to use Wittgenstein's expression which was introduced in section 4.2 above). This 'logical' or 'grammatical' difference, he expressed thus in his Cambridge lectures of 1931:

> a,b,c,- - -,and so on. Here - - - or 'and so on' stands for the rest of the alphabet, a definite number of letters.
>
> This is quite different from ⅓ = 0.33 - - - and so on. Here there is no definite number of digits, nor could there be for some superior being. The two examples have different grammars and rules. 0.33 - - - is not a makeshift: it has an exact grammar. (*LWL*, p. 90)[30]

In his 1939 lectures on the foundations of mathematics, Wittgenstein was still making the same point:

> There are two ways of using the expression 'and so on'. If I say, 'The alphabet is A, B, C, D, and so on', then 'and so on' is an abbreviation. But if I say, 'the cardinals are 1, 2, 3, 4, and so on', then it is not.—Hardy speaks as though it were always an abbreviation. As if a superman would write a huge series on a huge board—which is alright, but has nothing to do with the series of cardinals. (*LFM*, p. 255)[31]

A corollary of the thesis that infinity is 'the property of a law' and not of an extension is the idea, repeated many times,[32] that the infinite is not a number. Wittgenstein told his student John King, in 1930–1, that one of the mistakes he had made in the *Tractatus Logico-Philosophicus* was treating infinity as a number, and 'supposing that there can be an infinite number of propositions' (*LWL*, p. 119). This remark sheds light on an obscure corner of the *Tractatus Logico-Philosophicus*, since it implies (in support of Michael Wrigley's claim presented in section 2.2 above) that the setting in that book was infinite, i.e. that Wittgenstein assumed that there is an infinity of elementary propositions and that they can be joined together so that the quantifiers are indeed infinite conjunctions and disjunctions.

A consequence of the mistaken view that infinity is a number is the illusion that the infinite is to be compared with a finite quantity and then considered as a quantity—an enormous one—while 'it isn't itself a quantity' (*PR*, § 138).[33] It is worth noticing that Wittgenstein was

[30] See also *WVC*, pp. 203, 226, 228. [31] See also *RFM*, v, § 19.

[32] e.g. *WVC*, pp. 188, 226, 228; *PR*, §§ 138, 142, 172.

[33] See also *WVC*, p. 187; *RFM* v, §19; *LFM*, pp. 141–2.

certainly aware of the implications of this line of thought, as this remark from 1939 shows: 'If one were to justify a finitist position in mathematics, one should say just that in mathematics "infinite" does not mean anything huge. To say "There's nothing infinite" is in a sense nonsensical and ridiculous. But it *does* make sense to say we are not talking of anything huge here' (*LFM*, p. 255). This 'grammatical' point might appear to be nothing more than a rewording of the potentialist viewpoint (which is underlying the finitist account presented above in 4.2). There is something new, however, which sets Wittgenstein apart from Platonist, intuitionist, and strict finitist philosophies of mathematics. In order to present in a concise way what I think this difference consists of, I shall take as an example an infinite recursive sequence,

$$a_1, a_2, a_3, \ldots a_n$$

the Fibonacci sequence, already introduced in the previous section:

$$1, 1, 2, 3, 5, 8, 13, 21, 34, 55, 89, 144, 233, 377 \ldots$$

Recall that the Fibonacci numbers, are, for any $n \geq 2$, generated by the rule

$$a_n = a_{n-1} + a_{n-2}$$

with $a_1 = a_2 = 1$. One may distinguish two components in the rules for generating this sequence. First, a *local* component, which consists in an effective rule for calculating a_n, given a_{n-1} and a_{n-2}. Secondly, a *global* component, which would be the recognition that the entire collection is composed of nothing else but exactly what is given by such applications of the rule. The Platonist would certainly maintain that the whole infinite extension of the rule, or

$$1, 1, 2, 3, 5, 8, 13, 21, 34 \ldots a_{n-2}, a_{n-1}, a_n \ldots$$

is given to us once its local component is given. The intuitionist would beg to differ from the Platonist here by maintaining that the infinite extension cannot be regarded as ever completed, that it is only *in statu nascendi*. She would still maintain, however, that the local rule completely determines, in advance, a_{n-1} at any new step n. So we would have something like:

Step 1 1, 1, 2 . . .

. . .

Step n 1, 1, 2, 3, 5, 8, 13, 21, 34, . . . a_{n-2}, a_{n-1} . . .
Step $n + 1$ 1, 1, 2, 3, 5, 8, 13, 21, 34, . . . a_{n-2}, a_{n-1}, a_n . . .
. . .
ad infinitum

In contrast with the intuitionist, the strict finitist would make the further claim that there is no already defined a_{n-1} for a step n which has not been already computed. At each new step n, the community will decide what is the right a_{n-1}, and this still leaves a_n undecided.

Step 1 1, 1, 2, ?
. . .
Step n 1, 1, 2, 3, 5, 8, 13, 21, 34, . . . a_{n-2}, a_{n-1}, ?
Step $n + 1$ 1, 1, 2, 3, 5, 8, 13, 21, 34, . . . a_{n-2}, a_{n-1}, a_n , ?

This is precisely the kind of strict finitist position which has been associated with Wittgenstein in the past, notoriously by Michael Dummett in his review of the *Remarks on the Foundations of Mathematics* (Dummett 1978a: 248–68). One can see here the deep connection between strict finitism and the rule-following argument as it is understood by Saul Kripke (1982).[34] One finds a nice statement of this connection in Hilary Putnam's *Reason, Truth and History*:

> Whatever introspectible signs or 'presentations' I may be able to call up in connection with a concept cannot specify or constitute the *content* of the concept. Wittgenstein makes this point in a famous section which concerns 'following a rule'—say the rule 'add one'. Even if two species in two possible worlds (I state the argument in *most* un-Wittgensteinian terminology!) have the same mental signs in connection with the verbal formula 'add one', it is still possible that their *practice* might diverge; and it is the practice that fixes the interpretation: signs do not interpret themselves, as we saw. Even if someone pictures the relation 'A is the successor of B' (i.e. $A = B + 1$) just as we do and has agreed with us on some large finite set of cases [. . .] still he may have a divergent interpretation of 'successor' which will only reveal itself in future cases. [. . .]
>
> This has immediate relevance to the philosophy of mathematics, as well as to philosophy of language. First of all, there is the question of *finitism*: human practice, actual and potential, extends only finitely far. Even if we can

[34] I owe many of the preceding remarks to discussions with John Mayberry, but we differ on how to interpret Wittgenstein, since Mayberry would place him alongside strict finitists. This question will be taken up in Ch. 8. For an interesting use of an equivalent of the above 'local–global' distinction about rules, in the context of Wittgenstein's remarks on rule-following in the *Remarks on the Foundations of Mathematics*, see Floyd (1991: 161 f.).

say we can, we cannot 'go on counting forever'. If there are *possible divergent extensions of our practice, then there are possible divergent interpretations of even the natural number sequence*—our practice, or mental representations, etc., do not single out a unique 'standard model' of the natural number sequence. (Putnam 1981: 67)

(Putnam is speaking here about strict finitism, not finitism.) If there is any such thing, this is the received view about Wittgenstein on strict finitism and rule-following. I contend that this interpretation is incorrect, and shall come back to the issue of strict finitism at the end of this section and in Chapter 8.

Underlying the above three views is the common assumption of the existence of an extension. Under this assumption, the propositions appear as 'reports' about extensions and any disagreement is, so to speak, about the size of the extension. As I see it, Wittgenstein's originality consists precisely in rejecting this very assumption. (Incidentally, we are now in a position to understand Wittgenstein's requirement that there should be no 'opinions' in philosophical discussion. Indeed, any disagreement on the size of the extension would be, in Wittgenstein's jargon, the expression of an 'opinion'.)[35] Instead, with his injunction that in order to understand a mathematical proposition we ought to look 'at nothing but the proof' (*PR*, § 163; *PG*, pp. 369–70), Wittgenstein wanted to break the spell of the picture of an extension about which mathematical propositions are reports, in parallel with empirical statements which are reports about facts.

In this book I do not refrain from labelling Wittgenstein as a 'finitist'. But any use of the word 'finitism' to characterize his philosophy of mathematics is, in light of the preceding remarks, open to doubt. Moreover, he made a substantial number of remarks about 'finitism' (e.g. in his 1939 lectures, *LFM*, pp. 31, 111, 141, 255), which are more often than not of a negative character. These are very good reasons for believing that it is plainly misrepresenting Wittgenstein to call him a 'finitist'. But one ought not to conflate various meanings of the word 'finitism'. Reading carefully, one realizes that when he dismisses the so-called 'finitist' standpoint, he has almost invariably in mind a position which resembles that described above in the example of the Fibonacci sequence as 'strict finitism' (as does Putnam in the above quotation). As we shall see in a moment, Wittgenstein

[35] See Marion (1993) for comments to that effect.

rejected the argument about human epistemic limitations which underlies strict finitism. This is not what is meant here by calling him a 'finitist'. The example of the Fibonacci series helped us to see that Wittgenstein rejected an assumption underlying Platonism, intuitionism, and strict finitism, that of the existence of an extension about which propositions appear as 'reports' (in parallel with empirical statements). It is precisely this 'intensionalist' stance which is at the heart of Wittgenstein's position. Thus we see that, although he defended a potentialist account of infinity, his own 'grammatical' approach sets him apart from other potentialists such as Hilbertian finitists or modern-day intuitionists such as Dummett. Herein lies his originality. This point also coheres very well with the affinities between the reduction of arithmetic to the theory of 'operations' in the *Tractatus Logico-Philosophicus* and lambda-calculus, presented in section 2.1 above. Perhaps a paraphrase of one of Wittgenstein's own remarks about infinity (*RFM* ii, § 58) could be of help: 'Ought the word "finitism" to be avoided while describing Wittgenstein's philosophy of mathematics? Yes, where it appears to confer meaning upon his philosophy, instead of getting one from it.'

One crucial, distinctive aspect of Wittgenstein's finitism, which sets him apart from other finitist positions such as strict finitism, is his refusal to construe the issue about infinity in epistemological terms. As we saw in section 2.3, Ramsey claimed in 'Foundations of Mathematics' that there is nothing wrong with impredicativity, since one may refer to a man as the 'tallest in a group', even though that man would then be identified by means of a totality of which he is himself a member (Ramsey 1978: 192). Moreover, Ramsey extended this claim to the case where the totality in question is not merely finite, as in the case at hand, but infinite, thereby adopting an extreme form of Platonism about existence, according to which all properties and their extensions already exist, prior to our constructions or descriptions. The fact that some properties can only be described impredicatively is thus seen as caused by 'our inability to write propositions of infinite length', and Ramsey insisted that this inability is 'logically a mere accident' (1978: 192).

As far as I know, this is the first statement of an argument which has been used over and over by Platonists: the infinite powers of an omniscient being or God are not *logically* inconceivable. Bertrand Russell used it in a paper published in 1936, 'Limits of Empiricism', which he wrote in response to the 'finitist' arguments put forward by

Alice Ambrose in her paper 'Finitism in Mathematics'(1935), and by 'another finitist writer', Louis Goodstein, in 'an as yet unpublished paper' (Russell 1936: 144).[36] (An interesting coincidence: both Ambrose and Goodstein were students of Wittgenstein.) One finds echoes of Ramsey's argument when Russell claims, in his reply to Ambrose, that it may be 'medically' impossible but not 'logically' impossible to conceive of completed infinite processes:

> Miss Ambrose says it is *logically* impossible to run through the whole expansion of π. I should have said it was *medically* impossible. She thinks it *logically* impossible to know that there are not three consecutive 7's in π. But is it *logically* impossible that there should be an omniscient Deity? And if there is such a Deity, may he not reveal the answer to a mathematical Moses? And would not this be a demonstration? It seems to follow that, if a form of words p is syntactically correct, we always 'know what is meant by the statement that p is demonstrated.' If revelation is rejected as a demonstration, it will be found that we do not know of the existence of Cape Horn unless we have seen it. (Russell 1936: 143)

Wittgenstein knew Ambrose's paper quite well—he objected to its publication for reasons which he never made clear[37]—and Russell's, which he attacked frequently in his later writings (*RFM* iii, § 71; iv, § 27; vii, §§ 17, 21). One could adapt Russell's terminology and characterize the differences—Wittgenstein would say differences of 'opinion'—between strict finitism, intuitionism, and Platonism presented above with the help of Fibonacci numbers in terms of our 'medical' or epistemic limitations. On the one hand, an intuitionist would claim that humans are beings of finite powers who cannot complete an infinite process but can merely apply a rule for a finite number of steps. Strict finitists would distinguish themselves from intuitionists by insisting on putting an upper bound to the feasible applications of a rule. On the other hand, Platonists such as Russell would retort that these human limitations are only 'medical' impossibilities, and that we can, in turn, 'logically' conceive of completed infinite processes. Wittgenstein's position is better understood if we agree that episte-

[36] Goodstein's paper was probably 'Mathematical Systems', which appeared in *Mind* in 1939 (Goodstein 1965).

[37] Since Wittgenstein never made his reasons clear, either in print or in conversations with Moore or Ambrose, attempts at insinuating that Ambrose's use of some of Wittgenstein's ideas in her paper are misrepresenting his views are largely unsubstantiated. Alice Ambrose remains convinced to this day that she did not misinterpret Wittgenstein's ideas in her paper. But see Goodstein (1965: 80) for a comment in line with Wittgenstein.

mological arguments about our 'medical limitations' are in the domain of 'empirical' possibility, not of 'grammatical' possibility. His characterization of an infinite process as a possibility of the symbolism seems designed precisely to avoid arguing at the level of epistemology. Certainly, he did not deny that there are practical limitations to our ability to carry out a procedure. It would have been foolish to do so. But he deemed these 'empirical' limitations inessential:

> The rules for a number-system—say, the decimal system—contain everything that is infinite about the numbers. That, e.g. *these rules* set no limits on the left or right hand to the numerals; *this* is what contains the expression of infinity. Someone might perhaps say: True, but the numerals are still limited by their use and by writing materials and other factors. That is so, but that isn't expressed in the *rules* for their use, and it is only in these that their real essence is expressed. (*PR*, § 141)

Anticipating the discussion of strict finitism in Chapter 8, it must be said that this passage contains an important argument against the interpretation of Wittgenstein as a strict finitist: to regard limitations in our ability to write down numbers as essential, as the strict finitist would, was, according to him, to miss the point that these limitations are not expressed by the rules and that only what is expressed by the rules is the 'essence', so to speak. Strict finitist doubts can only occur if one confuses 'empirical' and 'grammatical' possibility. It is precisely because the 'and so on' is not to be understood as an abbreviation that we should not speak of an 'empirical' possibility or impossibility in connection with infinite processes, such as calculating the decimal expansion of 1/3, or that of π, as in this passage taken from the *Philosophical Investigations*: 'We should distinguish between the "and so on" which is, and the "and so on" which is not, an abbreviated notation. "And so on ad inf." is *not* such an abbreviation. The fact that we cannot write down all the digits of π is not an human shortcoming, as mathematicians sometimes think' (*PI*, § 208). If any more textual evidence is needed in support,[38] one can cite a passage from *Zettel* where Wittgenstein refers to the following remark by G. H. Hardy: 'That "the finite cannot understand the infinite" should surely be a theological and not a mathematical war-cry' (Hardy 1929: 5). Quoting it, Wittgenstein adds:

[38] See also *BB*, p. 54; *LFM*, pp. 31–2; *RFM* v, § 6.

True, the expression is inept. But what people are using it to try and say is: 'We mustn't have any juggling! How comes this leap from the finite to the infinite?' Nor is the expression all that nonsensical—only the 'finite' that can't conceive the infinite is not 'man' or 'our understanding' but the calculus. And *how* this conceives the infinite is well worth an investigation. (Z, § 273)[39]

Wittgenstein's characterization of the infinite as a 'grammatical' possibility was designed precisely to avoid misconstruing the nature of the infinite by getting entangled in an epistemological debate centred around what is possible for humans versus what is possible for God. In that sense, Wittgenstein's grammatical conception of the infinite bears no affinity with that of the strict finitist. It may even provide us with the basis for an argument against it.

In 'Wang's Paradox', Michael Dummett claimed that there is a danger for intuitionism in the fact that strict finitism renders its position unstable (Dummett 1978a: 249). Dummett's strategy against strict finitism is to point to the vagueness of predicates such as 'feasible' or 'surveyable' (when the strict finitist prescribes that numbers be feasible or that proofs be surveyable). This vagueness renders them susceptible to a variant of the Sorites.[40] For Dummett these expressions are for that reason semantically incoherent, and strict finitism, which admits of them, is not viable as a philosophy of mathematics (Dummett 1978a: 265). Dummett's strategy has a notorious flaw: the presumed incoherence of strict finitist notions is also a problem for intuitionism, since strict finitists make their case for stronger restrictions than those requested by intuitionists using radicalized versions of intuitionists' own arguments: if the strict finitist standpoint is in the end incoherent, there must be something wrong with the arguments themselves, i.e. with the very arguments needed to establish the intuitionist's case.[41] Wittgenstein's remarks on the infinite contain the germ of another argument, which avoids such unhappy consequences. The argument is, simply, that it is misleading to characterize the distinction between finite and infinite in terms of

[39] This passage originates in MS 116, § 54, dating from 1934.

[40] The predicate 'feasible' determines a weakly infinite and weakly finite totality. This totality forms a proper initial segment of the strongly infinite totality of the natural numbers, and one faces the following paradoxical situation: it is true that 0 is feasible, and if n is feasible, then Sn is also feasible, although the conclusion 'For every n, n is feasible' is false. This variant of the Sorites is dubbed 'Wang's paradox' by Dummett. More details will be given in section 8.1.

[41] Crispin Wright used this fact against Dummett's argument in Wright (1993).

abilities of human beings *vis-à-vis* those of God. In fact Wittgenstein would claim that this epistemological characterization comes to mind when one is not making the proper 'grammatical' distinction between finite and infinite series. If finitism (or intuitionism for that matter) is defended with the help of such an argument, then the epistemological debate is avoided and there is therefore no more risk of a radicalization of the arguments in the strict finitist direction. If this is true, Dummett's original worries would just disappear. Moreover, there is the possibility of criticizing both strict finitism and Platonism at the same time, because of their nonsensical construing of the distinction between finite and infinite series.

It is also worth noticing that Wittgenstein's argument is in many respects acceptable to the intuitionist; there is much similarity between Wittgenstein's and the intuitionist's conception of the infinite. Indeed, the intuitionist would say, following Dummett,[42] that to grasp an infinite structure is to grasp the process that generates it, while Wittgenstein claimed in his own terms that the infinite is 'the property of a law' (process), not of its extension. The intuitionist recognizes that an infinite structure is a process that will not terminate, while Wittgenstein spoke of a 'possibility of the symbolism', that of the fact that it sets no limits to its application. Moreover, we also saw that Wittgenstein insisted on not conceiving the infinite as a huge quantity. This is also an intuitionist argument: since Brouwer the intuitionists have accused the Platonist of illegitimately transferring a picture appropriate to the finite case to the infinite one (Brouwer 1967a: 336). Dummett does not say anything different:

> The Platonistic conception of an infinite structure as something which may be regarded both extensionally, that is, as the outcome of a process, and as a whole, that is, as if the process were completed, thus rests on a straightforward contradiction: an infinite process is spoken of as if it were merely a particularly long finite one. (1977: 57)

The distinction between the finite and the infinite is a conceptual one. It is only when it is recognized as such that it becomes irrelevant to refer to practical possibility, and it is no proper line of argument to invoke, as the Platonist does, a 'hypothetical being whose powers transcend our own' (Dummett 1977: 59). But we have here not just an argument against Platonism but also a point of contact between

[42] See Dummett (1977: 56).

intuitionism and Wittgenstein, and an argument undercutting strict finitism too. That it has not been recognized as such is to be explained by the fact that intuitionists have traditionally phrased their position in epistemological terms.

7

The Continuum

7.1. CAUCHY SEQUENCES AND THE DIAGONAL METHOD

Wittgenstein wrote or spoke extensively on the nature of the continuum during the years 1929–33. This is hardly surprising, since this topic was at the centre of the *Grundlagenstreit*: classical mathematicians felt that their beautiful Cantorian construction of the continuum was threatened by Brouwer's 'bolshevism'. Hilbert spoke of a 'putsch' and raised the spectre of Kronecker, the *Verbotsdiktator* (Hilbert 1935: 159–61). There are several methods in classical mathematics for introducing real numbers. Wittgenstein's comments were limited to the two principal ones, the introduction of the set of reals via Cauchy sequences of rationals and the method of Dedekind cuts. Wittgenstein's remarks on these are, again, of an elementary nature and of little interest to logicians today, but for us they are worth a closer look because they show much coherence in his commitment to finitism. In fact, it is only if we assume such a standpoint that these remarks make sense.

Cauchy reals are usually defined in the following manner: the sequence $(x_n) = (x_1, x_2, \ldots x_n, \ldots)$ is a Cauchy sequence if the difference of the terms x_p and x_q of the sequence (x_n), the number $|x_p - x_p|$, is getting closer to 0 as p and q are getting bigger, or

$$\lim_{p,\, q \to \infty} |x_p - x_q| = 0$$

The only necessary condition for a real sequence (x_n) to be convergent in **R** is that it is a Cauchy sequence. This is Cauchy's criterion of convergence: for all real $\varepsilon > 0$ there exists a natural $\mathrm{A} \in \mathbf{N}$ such that for any p, $q > \mathrm{A}$,

$$|x_p - x_q| < \varepsilon$$

This criterion plays a useful role since one can see that a sequence of rationals is convergent in **R** without the need to provide its limit.

With it we can define the following equality relation $=_r$ in the set of Cauchy sequences, here between (r_n) and (s_n):

$$(r_n) = r(s_n) \Leftrightarrow \lim_{n \to \infty} (r_n - s_n) = 0$$

This relation identifies sequences which differ in a finite initial part and also those such as $1/n$ and $1/n^2$, both converging to 0 but with different convergence behaviour. The set of reals is the set of equivalence classes of Cauchy sequences with respect to this equality relation. It is usually understood that the geometric line contains limits for all Cauchy sequences of rationals; therefore the set of reals is identified with the geometric continuum.

Both *Philosophical Remarks* and *Philosophical Grammar* contain extensive discussions of real numbers, whose purpose is, according to Lello Frascolla, 'to make clear which kind of law or prescription for the formation of convergent series of rationals can be legitimately considered a genuine generator of a real number' (1980b: 242). Wittgenstein formulated two requirements.[1] First—and this should come as no surprise—the prescription must be an effective rule:

(1) One must possess an effective rule to compute rational approximations.

This first requirement follows from his rejection of the extensionalist viewpoint. We saw in section 6.3 that according to Wittgenstein infinity is defined negatively by the fact that the law 'sets no limits' to its application (*PR*, § 139). Therefore an infinite sequence cannot be interpreted as already given in extension. Wittgenstein insisted that infinity should be seen as a 'possibility of the symbolism'. To quote him again: 'Infinity is the property of a law, not of its extension' (*LWL*, p. 13). Since Cauchy reals are defined as infinite sequences of rationals, they must be given by a law:

> A real number *yields* extensions, it is not an extension.
>
> A real number is: an arithmetical law which endlessly yields the places of a decimal fraction. (*PR*, § 186)

> The true nature of real numbers must be the induction. What I must look at in the real number, its sign, is the induction.—The '*So*' of which we may say 'and so on'. (*PR*, § 189)

> An irrational number is a process, not a result. (*AWL*, p. 221)

[1] I am here following the excellent work of Lello Frascolla (1980a: 666–7; 1980b: 242–3).

There is no room therefore for real numbers given in an extensional manner. This means that there is no room either for random real numbers, i.e. arbitrary decimal expansions. By definition, arbitrary infinite sequences are sequences generated not by a law but by an arbitrary selection of one term after another. The usual example of such arbitrary sequences is that of a binary decimal expansion whose digits are obtained by successive tosses of a coin. These arbitrary sequences are not in accordance with Wittgenstein's fundamental view of mathematics as being essentially a 'calculus', an activity of providing algorithms (see Chapter 1). Wittgenstein saw random real numbers as 'something empirical' (*WVC*, p. 83) and spoke in their case of an 'arithmetical experiment' (*PR*, §§ 190, 196).[2] He could not accept them as genuine mathematical objects, and clearly stated that reals should be defined in such a way—recursively—that one could not speak of an 'arithmetical experiment': 'Is an arithmetical experiment still possible when a recursive definition has been set up? I believe, obviously not; because via the recursion each stage becomes arithmetically comprehensible' (*PR*, § 194). So his continuum is composed only of reals generated by (arithmetical) laws:

> But can I be in doubt whether all the points of a line can actually be represented by arithmetical rules. Can I then ever find a point for which I can show that this is not the case? If it is given by means of a construction, then I can translate this into an arithmetical rule, and if it is given by chance, then there is, no matter how far I continue the approximation, an arithmetically defined decimal expansion which is concomitant with it. It is clear that a point corresponds to a rule. (*PR*, § 180)

This point has ramifications in Wittgenstein's criticism of the intuitionistic notion of choice sequences, which will be presented below. Although for some not all laws are recursive, it is clear that for him they are. So, as a result of his first requirement, his continuum boils down to the recursive continuum.

The second requirement on the formation of real numbers is that any real number must be effectively comparable with every rational number, since 'being comparable with other numbers is a fundamental characteristic of a number' (*PG*, p. 476):

[2] See *PR*, § 179, where Wittgenstein discusses the example of successive tosses of a coin as defining a point on the line by bisection and asks 'does this geometrical process define a *number*?' His answer is: 'But the operation is not an arithmetical one. (And the point which I call to my aid in my endless construction can't be given arithmetically at all).' See also *PR*, § 182.

It seems to be a good rule that what I will call a number is that which can be compared with any rational number taken at random. That is to say, that for which it can be established whether it is greater than, less than, or equal to a rational number.

That is to say, it makes sense to call a structure a number by analogy, if it is related to the rationals in ways which are analogous to (of the same multiplicity as) greater, less and equal to.

A real number is what can be compared with the rationals. (*PR*, § 191)[3]

This requirement originates in Wittgenstein's attempts at avoiding the extensionalist image of the line. Commenting on his insistence on effective comparability, Wittgenstein added:

I want to say that this is precisely what has been meant or looked for under the name 'irrational number'.

Indeed, the way the irrationals are introduced in text books always makes it sound as if what is being said is: Look, that isn't a rational number, but still there is a number there. But why then do we still call what *is* there a 'number'? And the answer must be: because there is a definite way for comparing it with the rational numbers. (*PR*, § 191)

The second requirement obviously eliminates numbers not given by law—random numbers—since 'you can compare a law with a law, but not a law with *no* law' (*PR*, § 181). More to the point, it also eliminates some numbers given by a law which fulfils the first requirement; it implies in particular the rejection of Brouwer's pendulum number, which is given recursively but not comparable in size to 0, from the domain of the reals. Wittgenstein was quite conscious of this fact:

The decisive thing about the construction of real numbers consists precisely in their comparability. It is only in virtue of this that the real numbers can be interpreted as points on a straight line.

If, now, there are constructions that cannot be compared with rational numbers, then we have no right to find them a place among the rational numbers. Thus they simply are not on the number lines. (In Brouwer it appears as if they were real numbers about which we merely did not *know* whether they were larger than, or smaller than, or equal to another rational number.) (*WVC*, p. 73)

This is the reason behind Wittgenstein's hostility towards Brouwer's pendulum number, for which it is impossible to tell whether it is bigger, equal, or smaller in relation to zero (see section 6.2 above). This

[3] See also *PR*, §§ 195–7; *WVC*, pp. 72–3.

second requirement is couched in vague terms, but the above remarks suggest, however, that the requirement is of the form:

(2) $\forall x, y \in \mathbf{R}\ (x > y \text{ or } x = y \text{ or } x < y)$

There are many constructive versions of the definition of the real numbers via the Cauchy sequences to be found in the literature. One notorious example is Errett Bishop's *Foundations of Constructive Analysis* (1967: ch. 2).[4] According to Bishop, a constructive real is a pair $((r_n), \mu)$, where (r_n) is a constructively given Cauchy sequence and μ a constructive rate-of-convergence function $\mu: \mathbf{N} \to \mathbf{N}$ such that:

$$\forall k > 0\ \forall n,m \geq \mu(k)\ \left[\,|r_n - r_m| < \frac{1}{k}\right]$$

Here, equality between reals is defined as:

$$((r_n), \mu) =_r ((s_n), \nu) \Leftrightarrow (r_n - s_n) \to 0$$

Real numbers *à la* Bishop clearly give some body to Wittgenstein's insistence, in (1), on effectivity. To use intuitionist terminology, Bishop is working in 'lawlike' analysis and his work corresponds to Brouwer's reconstruction of analysis without its idiosyncratic notion of choice sequences. There is, however, one noticeable difference, since statement (2) above, which is known as the Law of Trichotomy, entails a principle which is rejected by Bishop and his followers, the Limited Principle Omniscience.[5] There is therefore a tension between Wittgenstein's requirements which seems to have eluded him.

Since the publication of Bishop's book, many logicians have provided formal systems for his constructive analysis, such as Solomon Feferman's constructive theory of functions and classes T_0. In Feferman's T_0, there seems no clear way of distinguishing between 'real' and 'computable real' numbers, although there are different notions of recursivity, one of them corresponding to Bishop's (Feferman 1984: 152). There is also another school of constructivism which limits itself to 'recursive' or 'computable' analysis.[6] The rather

[4] See also the revised edition, Bishop and Bridges (1985: ch. 2).

[5] This fact was overlooked in Marion (1995c). For statements of Limited Principle Omniscience, see Bishop (1967: 9; 1975: 511); Bishop and Bridges (1985: 11). For a discussion of the relation of this principle to the Law of Trichotomy, see Bridges (1994: 33, 45).

[6] I see no need to introduce here the Russian constructivist school of Shanin and Markov. (For details on the relations between intuitionism, Bishop, and Russian constructivism, see Bridges and Richman (1987).) A good representative of 'computable'

vague notions of 'law' or 'rule' which Wittgenstein made ample use of are replaced here by the precisely defined concept of recursive function. All 'objects' are given by numbers and these are manipulated only by recursive functions. The connections with Wittgenstein's requirements for the formation of real numbers are obvious, since both Wittgenstein and representatives of recursive analysis require an effective method of computation of the approximated rational values.[7] But, again, statement (2) would not hold.

An obvious objection to Wittgenstein's requirements (1) and (2) is that they force us to adopt a unduly restricted model of the continuum. Wittgenstein had this answer:

> 'The decimal fractions developed in accordance with a law still need supplementing by an infinite set of irregular infinite decimal fractions that would be "brushed under the carpet" if we were to *restrict* ourselves to those *generated by a law*.' Where is there such an infinite decimal that is generated by no law? And how would we notice that it was missing? Where is the gap it is needed to feel? (*PG*, p. 473)

Once again, we see here Wittgenstein's deeply anti-Platonist outlook: there is nothing pre-existing to our constructions, therefore no gap to be noticed. In this passage of the *Philosophical Grammar*, Wittgenstein developed his argument along the following line: since there is no point corresponding to an irrational, we only have an infinite series of approximations given by a law. If we stop at any point in the process of infinite approximation, then there is another series coinciding with it. Hence no gap. Here, Wittgenstein's example is π:

> If an irrational number is given through the totality of its approximations, then up to *any* point taken at random there is a series coinciding with that

analysis is Oliver Aberth. See Aberth (1970; 1980) and Beeson (1985: ch. 4). Aberth's work does not represent the Russian constructivist school. The opening paragraph of Aberth's 'Computable Analysis and Differential Equations' is a statement of his programme: 'Computable analysis is an analysis restricted to the field of computable numbers where a real number is called computable if there is an algorithm for obtaining precise rational approximations. Algorithms also are the basis for the definitions of the functions and sequences of computable analysis, and in every instance of a number, function or sequence, an appropriate defining algorithm is assumed available' (Aberth 1970: 47). As with Wittgenstein, Aberth requires an effective method of computation of the approximated rational values. Moreover, Aberth's formulation also makes obvious the central role of algorithms, a role which corresponds perfectly to Wittgenstein's claim (see Ch. 1) that mathematics is essentially of algorithmic nature.

[7] It is fitting to notice in this context the pioneering role played in the development of recursive analysis by Wittgenstein's student Louis Goodstein. See Goodstein (1957; 1961).

> of π. Admittedly for each such series there is a point where they diverge. But this point can lie arbitrarily far 'out', so that for any series agreeing with π I can find one agreeing with it still further. And so if I have the totality of all irrational numbers except π, and now insert π I cannot cite a point at which π is now really needed. At *every* point it has a companion agreeing with it from the beginning on. (*PG*, p. 473)

Whatever worth this sort of argument has, we ought simply to notice again that it is rooted in Wittgenstein's strong, peculiar anti-Platonist stance, according to which there are no infinite extensions and only laws 'reach to infinity': 'If from the very outset only laws reach to infinity, the question whether the totality of laws exhaust the totality of infinite decimal fractions can make no sense at all' (*PR*, § 181).

It is difficult not to say a few words in this chapter on Wittgenstein's remarks on Cantor's *Diagonalverfahren* or method of the diagonal, and its use in the proof of the uncountability or non-denumerability of the set of all real numbers. There are very few remarks on this topic in the early transitional writings (e.g. *WA* i, p. 98). The most substantial remarks are found in part ii of the *Remarks on the Foundations of Mathematics*, which was written *c.*1938. For that reason I ought not to mention the topic here, but it can hardly be avoided in a discussion on real numbers and the continuum. Moreover, Wittgenstein's remarks on this topic are, as we shall see, quite in line with the views presented so far in this chapter.

Cantor used his diagonal method in one of his most important results, which states that the set of all real numbers in the interval $0 \leq x \leq 1$ is uncountable. Since the rationals are countable and the union of any pair of countable disjoint sets is countable, Cantor's proof had to rely on the uncountability of the irrationals. These are represented as an array of infinite (non-recurring) decimals:

$$
\begin{array}{lllllll}
\mathrm{a}\,. & \mathrm{a}_1 & \mathrm{a}_2 & \mathrm{a}_3 & \mathrm{a}_4 & \mathrm{a}_5 & \ldots\ldots \\
\mathrm{b}\,. & \mathrm{b}_1 & \mathrm{b}_2 & \mathrm{b}_3 & \mathrm{b}_4 & \mathrm{b}_5 & \ldots\ldots \\
\mathrm{c}\,. & \mathrm{c}_1 & \mathrm{c}_2 & \mathrm{c}_3 & \mathrm{c}_4 & \mathrm{c}_5 & \ldots\ldots \\
\mathrm{d}\,. & \mathrm{d}_1 & \mathrm{d}_2 & \mathrm{d}_3 & \mathrm{d}_4 & \mathrm{d}_5 & \ldots\ldots \\
\mathrm{e}\,. & \mathrm{e}_1 & \mathrm{e}_2 & \mathrm{e}_3 & \mathrm{e}_4 & \mathrm{e}_5 & \ldots\ldots \\
\ldots & & & & & &
\end{array}
$$

(Here letters on the left of the point represent a numeral, and those on the right any digit from 0 to 9.) Cantor observed that the array cannot be complete, since one can define a decimal which differs at the *n*th place from each *n*th decimal in the array. One finds very harsh words on Cantor's proof in the *Remarks on the Foundations of Mathematics*, for example Wittgenstein's description of it as 'a puffed-up proof' (*RFM* ii, § 21) and as '*Hokus Pokus*' (§ 22). A closer look shows that Wittgenstein did not oppose the diagonal method itself, but the interpretation of the result of its application in Cantor's proof. It is difficult indeed to oppose the diagonal method itself, even from a constructivist point of view. It could be described as a rule to operate on rules, and as such it makes no commitment to completed infinities. Indeed, while Cantor simply observes that there is a decimal that differs in the *n*th place from each *n*th decimal in the array given above, one can, as Wittgenstein presumably would, restrict the diagonal to effectively computable infinite decimals, i.e. to decimals whose every *n*th place can be calculated. If for each *n* one can effectively determine what the *n*th decimal in the array should be, then one can effectively compute a decimal which is not part of the array. Hence the method becomes purely constructive. This is how Wittgenstein understood the diagonal: he spoke of it as showing that 'it makes no sense to talk about a "series of all real numbers" ' (*RFM* ii, § 16). This is precisely what he saw as the 'proper purpose' of the diagonal:

> Surely—if anyone tried day-in day-out 'to put all irrational numbers into a series' we could say: 'Leave it alone; it means nothing; don't you see, if you establish a series, I should come along with the diagonal series!' This might get him to abandon his undertaking. Well, that would be useful. And it strikes me as if this were the whole and proper purpose of this method. (§ 13)

Looking at the diagonal as a rule to operate on rules, the result is not about extensions. For Wittgenstein there are no such things as extensions to be compared here, and Cantor's proof is misinterpreted:

> one pretends to compare the 'set' of real numbers in magnitude with that of the cardinal numbers. The difference in kind between the two conceptions is represented, by a skew form of expression, as difference of extension. I believe, and hope, that a future generation will laugh at this hocus pocus. (§ 22)

The proof is supposed to provide the meaning; one ought to look at the proof instead of getting confused by the verbal expression of its result:

> The result of a calculation expressed verbally is to be regarded with suspicion. The *calculation* illuminates the meaning of the expression in words. It

is the *finer* instrument for determining the meaning. If you want to know what the verbal expression means, look at the calculation; not the other way about. The verbal expression casts only a dim general glow over the calculation: but a calculation a brilliant light on the verbal expression. (§ 7)

Can the application of the diagonal method in Cantor's proof be said to prove the uncountability of the reals? For Wittgenstein the diagonal only gives a 'rule for step-by-step construction of numbers that are successively different' from a given sequence of them (*RFM* ii, § 3). His imaginary opponent would retort that 'your rule already reaches to infinity, so you already know quite precisely that the diagonal series will be different from any other!' (§ 9). But this is only relative to one enumeration. Let us suppose that we have a formal system within which real numbers are denumerable. Cantor's proof shows us that the function which enumerates real numbers is itself not definable in the system. One may thus refuse to recognize that Cantor's proof provides us with an absolute proof of non-denumerability. This reasoning seems to me to reflect Wittgenstein's attitude towards the diagonal. This attitude is reminiscent of that of Brouwer in one of his earliest papers, 'Die möglichen Mächtigkeiten' (1975: 102–4).[8] As is typical of him, Brouwer opened his paper with a mention of the *Ur-intuition der Zweieinigkeit* or fundamental intuition of bi-unity (p. 102), from which one could obtain the order type of natural numbers by iterating 'next' and the order type of rational numbers by iterating 'between'. The 'powers' thus obtained are only countable, however, and Brouwer claimed in the paper that there are two ways of obtaining higher powers: the diagonal method and the construction of the continuum. But he claimed that the possibility of constructing newer ordinals through the diagonal method does not show the uncountability of the set of countable ordinals, and he thought that this shows that the set of such ordinals (known to Cantorians as the 'second number class') does not exist (Brouwer

[8] Cf. also the argument put forth by Alistair Watson (1938: 448–9). Watson uses the impossibility of applying the diagonal to Turing machines to show that 'in no way can we arrive at anything in the way of a transfinite number' (p. 449). It was Watson who introduced Turing to Wittgenstein, and it is noticeable that Wittgenstein, contrary to famous cases involving Braithwaite and Ambrose, did not object to Watson's recognition that the views expressed in his paper owe much to 'lengthy discussions with a number of people, especially Mr. Turing and Dr. Wittgenstein of Cambridge' (p. 445)! It was also during the year 1938 that Wittgenstein wrote the remarks on Cantor's proof quoted above. For a more recent discussion of these from a perspective akin to Wittgenstein's, see Wright (1985).

1975: 102–3).[9] I do not see that Wittgenstein was saying anything different in 1938.

7.2. CHOICE SEQUENCES AND DEDEKIND CUTS

Before discussing Dedekind cuts, we should have a quick look at Wittgenstein's remarks on intuitionistic choices sequences. Wittgenstein's criticisms of these two constructions are similar in some respect. To grasp the idea of choice sequences, it is useful to represent real numbers by the bisection of the line. One would then have an infinitely branching binary tree (Fig. 2).One can then designate points by binary fractions, and since in the continuum no boundaries can be set to this subdivision, it continues ever more precisely defining earlier points. The tree will have non-denumerably many paths, but only denumerably many nodes. Some, but not all, paths down the tree are regular and can be defined by a law. These paths are what intuitionists call 'lawlike' sequences, and they are captured by recursive models of the continuum. Intuitionism distinguishes itself from such programmes by providing, alongside a theory of recursive real numbers, a theory about sequences in which there are also free selections.[10] These are either 'choice' or 'lawless' sequences. In the current intuitionistic literature, a lawless sequence is defined as an expansion not defined by any law. At any stage the only knowledge available is about values previously obtained, and no information about future stages is available. The example of a sequence resulting from successive tosses of a coin or casts of a die is a good analogy here (Troelstra 1977: 12). A choice sequence is somewhat an intermediary between lawlike and lawless sequences.

[9] As for the continuum, Brouwer likened the real line to the set of all paths down an infinite binary tree and claimed that the continuum is not properly speaking a set, but a 'matrix' (Brouwer 1975: 103).

[10] Michael Dummett provided the following justification: 'Intuitionism aims, however, to reform mathematics, not to prune it; according to it, scarcely any of the ideas of classical mathematics is wholly spurious, but all are deformed by being systematically misconstrued. Hence intuitionism retains both fundamental ideas which go to form the classical continuum, admitting not only infinite sequences determined in advance by an effective rule for computing their terms, but also ones in whose generation free selection plays a part' (1977: 62). Bishop simply rejected the theory of choice sequences because he thought that it 'makes mathematics so unpalatable to mathematicians, and foredooms the whole of Brouwer's program' (1967: 6).

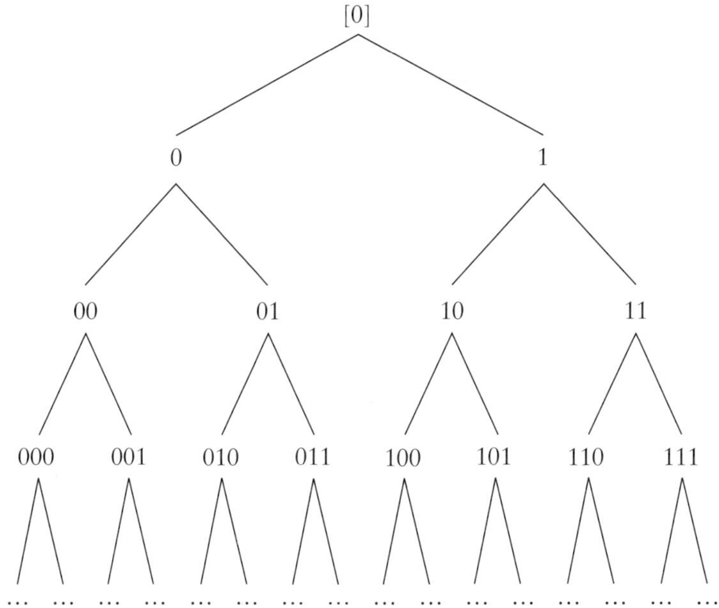

The notion of choice sequences was first introduced by Émile Borel in his lecture to the International Congress of Mathematicians in 1908, for reasons linked with his work on probability. One has to recall here Borel's views on existence (see Chapter 1 and section 3.3 above): he was moving from the view according to which the only admissible objects are 'defined in a finite number of words', within which there is little room for chance, to a more liberal attitude according to which a denumerable infinity of successive and arbitrary choices is allowed. Thus non-computable functions, instead of being disallowed, are subsumed under the calculus of probability. According to Borel, real numbers, when they exist, are computable, and Cantor's proof tells us the continuum cannot be exhausted by a list of computable reals. But the continuum is thus only a 'purely negative notion', a 'transitory instrument' in the study of denumerable sets (Borel 1972: 1055–6). In probability theory Borel's attitude was translated into the view, presented in 'Les Probabilités dénombrables et leurs applications mathématiques', that a third form of probability, which he called 'denumerable probability' (*probabilités dénombrables*), should be inserted between 'discrete probabilities' and 'continuous probabilities' (p. 1055). This philosophical move is not without crucial mathematical consequences, since it is in this very paper that Borel obtained his strong law of large numbers, the first result of modern probability theory.[11]

In his doctoral thesis Brouwer also conceived of the continuum as a 'purely negative notion', but he initially rejected choice sequences because of his firm conviction, at the time, that real numbers had to be identified with a law. At first, he could not admit convergent choice sequences as sufficiently well defined to be arguments of real functions.[12] Wittgenstein's standpoint, as presented above, has affinities with Brouwer's early views. But Brouwer was already showing signs in 1912 of admitting choice sequences, something Wittgenstein never did. In his inaugural address, 'Intuitionism and Formalism', Brouwer presented choice sequences as an empty 'formalist' creation:

> Let us consider the concept: 'real number between 0 and 1'. For the formalist this conception is equivalent to 'elementary series of digits after the decimal point', for the intuitionist it means 'law for the construction of an elementary series of digits after the decimal point, built up by means of a finite number of operations'. And when the formalist creates the 'set of all real numbers between 0 and 1', these words are without meaning for the intuitionist, even whether one thinks of the real numbers of the formalist, determined by elementary series of freely selected digits, or of the real numbers of the intuitionist, determined by finite laws of construction. (Brouwer 1975: 134)

But in the same passage he opened the door: 'for the intuitionist can only construct denumerable sets of mathematical objects and if, on the basis of the intuition of the linear continuum, he admits elementary series of free selections' (pp. 134–5). As soon as 1914 Brouwer liberalized his notion of a constructive procedure in order to include choices and not just finite laws of generation:

> for the intuitionist only well-constructed infinite sets exist, which are assembled from a part of the first kind, constructed as a single fundamental sequence, and a part of the second kind, based on a fundamental sequence *f* conceived as a Fréchet V-class, whose elements are determined by a sequence of choices from elements of a finite set or of a fundamental sequence in such a way that to every sequence of choices corresponds nested intervals of *f* whose widths converge to zero, and that there exists final segments of any

[11] For an excellent account of the development of modern probability theory which is sensitive to Borel's constructivism, see von Plato (1993).

[12] On this point see van Stigt (1990: 358–9).

two sequences of intervals, which are corresponding to any two sequences of choices, lying apart from each other. (Brouwer 1975: 140)

Brouwer became aware of the implications for logic of his new notion. In 1917–18 he introduced his concept of 'spread' and began exploiting choice sequences, introducing his first continuity arguments.[13] Choice sequences thus acquired a new status. One may here follow Kreisel and look at intuitionistic logic as an extension of Fregean logical foundations, i.e. as the logic of incompletely defined objects such as choice sequences (Kreisel 1984: 72). Although Brouwer devoted the bulk of his logical studies to this topic, one had to wait for the work of Georg Kreisel and Anne Troelstra for a more extensive development of the theory of lawless and choice sequences (Kreisel 1968; Troelstra 1977).

To come back to the 1920s, Weyl paid tribute in 'Über die neue Grundlagenkrise der Mathematik' to Brouwer for showing that number sequences created through a 'free act of choice' (*freie Wahlakte*) could be objects of mathematical conceptualization (Weyl 1968, ii. 152–3). Following Brouwer, Weyl emphasized the importance of the *Wahlfolge*: 'As the law φ determining an infinite sequence represents the isolated real number, so is the choice sequence limited by no laws in the freedom of its expansion representing the continuum' (p. 153). Later, in 'Die heutige Erkenntnislage in der Mathematik', Weyl insisted that the most important characteristic mark of 'choice sequences in becoming' (*werdenden Wahlfolgen*) is that they are closed under continuous operations:

That it is mathematically possible to operate with choice sequences in becoming, is adequately illustrated by the fact that one can provide an ordering between the choice sequences. E. g. the formula

$$n_h = m_1 + m_2 + \ldots + m_h \quad (h = 1, 2, 3, \ldots)$$

contains a law according to which a number sequence in becoming n_1, n_2, n_3, . . . is built from a sequence in becoming m_1, m_2, m_3, . . . given by free acts of choice, their coming into being keeping at the same pace. (Weyl 1968: ii. 531)

The gist of Weyl's remark is the fact that it is possible to carry out lawlike operations on sequences whose elements need not be determined by a law.[14] Wittgenstein probably learned about choice sequences from reading these very papers by Weyl; in fact, there is

[13] See Troelstra (1982: 470–2). [14] See Troelstra (1977: 12–13).

what seems to be a direct reference to the above remark in one of the discussions with Schlick and Waismann:

> A freely developing sequence is in the first place something empirical. It is nothing but the numbers that I write down on paper. If Weyl believes that it is a mathematical structure because I can derive a freely developing sequence from another one by means of a general law, e.g.,
>
> $m_1, m_2, m_3, \ldots$
>
> $m_1, m_1 + m_2, m_1 + m_3, \ldots$
>
> then the following is to be said against it: No, this shows only that I can add numbers, but not that a freely developing sequence is an admissible mathematical concept. (*WVC*, p. 83)

Here, Wittgenstein refuses to admit choice sequences as bona fide mathematical objects, implying that the fact that one can operate with choice sequences is not a sufficient factor for their admissibility as such. In his remarks, Wittgenstein conflated lawless and choice sequences—this is perfectly understandable, since the notions were not clearly distinguished in those days; the notion of 'absolutely free choice sequence' was introduced by Kreisel in 1958 (1958c)—and concentrated on the examples of sequences resulting from successive tosses of a coin or casts of a die. Such sequences, according to Wittgenstein, are given only through a description:

> the description 'endless process of choosing between 1 and 0' does not determine a law in the writing of a decimal. Perhaps you feel like saying: the prescription for endless choice between 0 and 1 in this case would be a symbol like '$0\,{}^{000}_{111}$ . . . ad. inf.'. But if an adumbrate a law thus '0.001001001 . . . ad. inf.', what I want to show is not the finite section of the series as a specimen of the infinite series, but rather the kind of regularity to be perceived in it. But in '$0\,{}^{000}_{111}$ . . . ad. inf.' I don't perceive *any* law,—on the contrary, precisely that a law is absent. (*PG*, p. 472)[15]

And, of course, a description '*is* not arithmetic' (*WVC*, p. 83), it is law*less*, hence unacceptable. The very idea of such a description, i.e. the idea of an 'arithmetical experiment', was seen as absurd: 'A number as the result of an arithmetical experiment, and so the description of a number, is an absurdity. The experiment would be the description, not the representation of a number' (*PR* § 196). Wittgenstein's stubborn denial is rooted in his deep conviction, described in section 6.3 above, that only laws 'reach to infinity'. But

[15] Cf. *AWL*, p. 221.

he does not provide any argument against the admissibility of choice sequences as mathematical objects. At any rate, his negative comments are beside the point. Any intuitionist would agree that any particular choice sequence lacks the mathematical existence of a law. The point of the exercise does not consist in justifying particular choice sequences as mathematical objects but, rather, to provide a general schema which would be studied for its own sake.[16] So Wittgenstein's criticisms of intuitionism are once again found to be unconvincing. Be that as it may, his rejection of such a distinguishing figure of intuitionism as the theory of choice sequences is of interest to us, as it indicates clearly that we should separate him definitely from Brouwer and place him firmly in the tradition of Kronecker, Skolem, and Bishop.

The remarks on the notion of Dedekind cut contain no new element to be added to Wittgenstein's conception of the continuum as presented above. In fact, the criticisms he levelled against it bear much resemblance to criticisms of choice sequences we have just seen. The very few assessments of these remarks to be found in the secondary literature, such as that of Paul Bernays in his review of the *Remarks on the Foundations of Mathematics* (Bernays 1986: 175), are usually negative. I shall nevertheless give a brief presentation in order to clarify their gist and to show that the points Wittgenstein was trying to get across are rather uncontroversial. The problem with these remarks is simply that in the end one cannot extract from them an original, convincing argument against the Platonist picture of the real line, which he thought was vitiating Dedekind's construction. At any rate, Wittgenstein's criticisms lose their value when we realize how much he liked the definition of reals via Cauchy sequences, which we know now to be (classically) equivalent to Dedekind cuts.[17] Moreover, there are now constructive versions of the Dedekind cuts (Troelstra and van Dalen 1988: 270–7). One is left to wonder what Wittgenstein would have thought of them.

The point of departure of Dedekind's work on irrationals was his observation that in every division of the line in two classes of points such that every point in one class is to the left of the second class,

[16] This point is made in Frascolla (1980a: 673).

[17] The proof that both structures, Cauchy and Dedekind reals, are isomorphic needs the choice principle AC_{00} which is admissible to Wittgenstein (see section 3.3 above). Even if it is rejected, it is said that in the models where it fails the Dedekind reals prove to be more interesting (Troelstra and van Dalen 1988: 251, 270 f.).

there is only one point that produces the division. This is the essence of continuity:

> If all points of the straight line fall into two classes such that every point of the first class lies to the left of every point of the second class, then there exists one and only one point which produces this division of all points into two classes, thus severing of the straight line into two portions. (Dedekind 1963a: p. 11)

Dedekind carried over this image to **Q**, calling a 'cut' (*Schnitt*) any division of the rational numbers into two classes such that any number in the first class is less than any number in the second. For classes A_1 and A_2, the cut is denoted by (A_1, A_2) (Dedekind 1963a: 12–13). There are two kinds of cut in the rational line. First, there are cuts determined by a rational number. In these cases there is either a largest number in A_1 or a smallest number in A_2 and, conversely, any cut in the rationals where there is a largest number in A_1 or a smallest one in A_2 is determined by a rational number. Secondly, there are cuts not produced by rationals but by irrational numbers such as $\sqrt{2}$. Here, Dedekind, who thought that 'numbers are free creations of the human mind' (1963b: 31–2), claimed that we 'create' the irrational number: 'Whenever, then, we have to do with a cut (A_1, A_2) produced by no rational number, we create a new, an *irrational* number α, which we regard as completely defined by this cut (A_1, A_2); we shall say that the number α corresponds to this cut, or that it produces this cut' (1963a: 15). The rhetoric about 'free creation' of irrational numbers should not confuse anyone about Dedekind's Platonism: it is the real number, whatever it is, which 'corresponds to' or 'produces' a cut (Dedekind 1963: 15),[18] and it is *represented* by the set of all rational numbers that it is greater (or smaller) than. Dedekind carried on by pointing out that numbers such as α possess the following three properties:

1) If $\alpha > \beta$ and $\beta > \gamma$, then $\alpha > \gamma$.
2) There are an infinite number of different numbers between the real numbers α and β.
3) If α is a real, then the real numbers are divided in two classes A_1 and A_2, each of which containing an infinity of numbers and each member of A_1 is less than α and each member of A_2 is greater than α.

[18] For Dedekind's objections to Weber's suggestion that irrational numbers should be taken *to be* the cuts, see his letter to Weber, 24 Jan. 1888 (Dedekind 1932: iii. 489).

This is how Dedekind shows that the real numbers possess continuity, i.e. that if **R** is divided into two classes A_1 and A_2 such that each member of A_1 is less than all members of A_2, then there is one and only one number α which divides **R** at this point.

In Wittgenstein's milieu, it was *de bon ton* to criticize Dedekind's theory. It was criticized by Frege (1980b: § 139–40) and by Russell, who spoke of Dedekind's method as having 'the same advantages of theft over honest toil' (1919: 71). In the same vein, Wittgenstein had only critical comments about Dedekind cuts. This is one of the topics in the foundations of mathematics on which he never changed his mind. The ideas developed in the early 1930s are still unchanged in the mid-1940s. I take it that he was trying to get across two points. Dedekind viewed a cut as a purely arithmetic phenomenon (Dedekind 1963b: 37). But everyone now knows that this is true only if arithmetic is considered as part of logic, and set-theoretic principles are disguised as logical ones. Wittgenstein would claim here—this is the first point—that the cut is not an arithmetical operation. The reasoning here is similar to that which led him to reject intuitionist choice sequences. Dedekind's theory allows for cuts representing numbers not given by an arithmetical law, and that was unacceptable to Wittgenstein:

> We should make a distinction between a class of tosses, or mere choices, and a way, or rule, for making choices. The latter defines an irrational number. An irrational number is a procees, not a result. We have a tendency to think that there is one result produced by $\sqrt{2}$, viz., an infinite decimal fraction. $\sqrt{2}$ produces a series of results, but no single result. $\sqrt{2}$ is a *rule* for producing a fraction, not an extension. Now there is this difference between the rule for constructing a decimal by repeated throws of a coin and the rule for working out places of $\sqrt{2}$, namely, that we have a fixed method for deciding for any rational number whether it is larger or smaller than $\sqrt{2}$. (*AWL*, p. 221)

At the heart of Wittgenstein's objections to Dedekind cuts was his conviction that it is a 'pernicious idiom of set theory' (*PR*, § 173) that we say that the ultimate parts of a line are points. The cut is 'an extensional image' (*RFM* v, § 34), and the idea that there is a point lying at the cut is 'a misleading picture' (*PG*, p. 472).[19] This comes

[19] It is worth noticing here that this was also Weyl's opinion: 'the concept of point must be considered as a limit-idea, "point" is the representation of the *limit* of a division continued to infinity' (1968: ii. 177). Probably influenced by Wittgenstein, Waismann also made some similar remarks in his *Introduction to Mathematical Thinking* (1951: 192).

out clearly, in relation to the example of the toss of a coin, in his 1932–3 lectures, 'Philosophy for Mathematicians':

> Suppose we divide a line *AD* in accordance with the rule of bisecting the interval on the left if we throw tails and on the right if we throw heads. For example,
>
> *A*——————————|————|—|—|———|————————*D*
>
> Here we believe ourselves to be determining a point by ever decreasing intervals in which it lives by repeatedly throwing the coin and thereby always diminishing the abode of the point. Also, we take it that the point corresponds to an irregular infinite decimal. But by throwing a coin so as to decrease the interval, we have not determined a point endlessly approached by the cuts made in accordance with the repeated tosses of the coin. We have really a series of intervals, which will always remain such. After every throw the point is still infinitely indeterminate. The trouble is with our imagery. (*AWL*, p. 200)[20]

As we already saw, for Wittgenstein the procedure of determining the whereabouts of a point on the line by successive tosses of a coin or throwing of a die, as in the following passage, was unacceptable, simply because the process is not regulated by a law:

> It is not enough that someone should—supposedly—determine a point ever more closely by narrowing down its whereabouts. We must be able to construct *it*. To be sure, continued throwing of a die indefinitely restricts the possible whereabouts of a point, but it doesn't determine a point. After *every* throw (or every choice) the point is still infinitely indeterminate—or, more correctly, after every throw it is infinitely indeterminate. (*PG*, p. 477)

But this not all: the very idea that there is something, a point, that corresponds to the completed process of approximation of $\sqrt{2}$ or π (and here there are rules) is misleading. We are here misled by the image of the line composed of points, pre-existing our constructions. For, according to Wittgenstein, there are no such points, gaps, etc. We only have our rules of approximation. As he would say, 'the rule for working out places of $\sqrt{2}$ is itself the numeral for the irrational number' or, to speak in dangerously misleading geometrical terms, 'mathematical *rules* are the points' (*PG*, p. 484). In this he took the opposite stance to Dedekind, for whom there are such points (it is the irrational, already there, which 'produces' the cut). This leads us to the second point: for Wittgenstein the definition of real numbers

[20] Cf. *PR*, § 179 and *PG*, p. 484.

via cuts is evidently geometrically motivated.[21] As Wittgenstein understood Dedekind, if we cut the line indiscriminately, we would find either a rational number or an irrational one, as if these were already on the line independently of our calculations. But he thought it a mistake to think that a real number could help one to do the cut, as if it were already given with the extensional representation of the straight line. He profoundly disliked this Platonist imagery, and displayed his disapproval in remarks written as late as the mid-1940s: 'The misleading thing about Dedekind's conception is the idea that the real numbers are spread out in the number line. They may be known or not; that doesn't matter. And in this way all that one needs to do is to cut or divide into classes, and one has dealt with them all' (*RFM* v, § 36). These criticisms can be said to be unfair to Dedekind, who claimed that, if geometrical intuition is 'exceedingly useful, from the didactic standpoint', it can on the other hand 'make no claim to be scientific', and who thought that he had found a 'purely arithmetic and perfectly rigorous foundation for the principles of infinitesimal analysis' (Dedekind 1963a: 2–3). That, in spite of his avowed intentions, Dedekind's definition of real numbers as cuts was nevertheless geometrically motivated is best seen from the following quotation:

> Of the greatest importance, however, is the fact that in the straight line L there are infinitely many points which correspond to no rational numbers. . . If now as is our desire, we try to follow up arithmetically all phenomena in the straight line, the domain of rational numbers is insufficient and it becomes absolutely necessary that the instrument **R** constructed by the creation of the rational numbers be essentially improved by the creation of new numbers such that the domain of numbers shall gain the same completeness, or as we may say at once, the same *continuity*, as the straight line. (Dedekind 1963a: 8–9)

Dedekind's intention was to complete the field of rational numbers Q to a 'continuous manifold' in order to pursue the properties of the straight line arithmetically. This is done by carrying the geometric continuity principle over to arithmetic and by letting the real numbers be cuts of rationals. Then Dedekind proved that the system of all cuts of rationals is continuous (Theorem IV, Dedekind 1963a: 20). For Wittgenstein, the proof of Dedekind's theorem seemed to be justified by an image while, on the contrary, it is the image that should be justified by the proof.

[21] This seems to be also Russell's opinion (1919: 71).

In his review of the *Remarks on the Foundations of Mathematics*, Paul Bernays pointed out that Dedekind's theory might at first appear unsatisfactory because when one defines cuts as sets of numbers and then tries to give the arithmetical law of such sets, the very notion of 'law' becomes vague. Weyl has shown that here one faces the problem of the *circulus vitiosus* (1987: 109–17). But impredicativity causes no qualms to those adhering steadfastly to the extensional viewpoint (see section 2.3 above), as Bernays pointed out: 'difficulties are not encountered if the extensional standpoint is consistently retained, and Dedekind's conception can certainly be understood in this sense and was probably so understood by Dedekind himself' (Bernays 1986: 175).[22] For that reason, no Platonist will be impressed by Wittgenstein's remarks. It remains true, however, that they are based on the relatively uncontroversial facts that Dedekind's approach was geometrically motivated and that it was set-theoretical. They also show his deep distrust of the Platonist imagery and cohere very well with his other remarks on the continuum. As for those who sneer at Wittgenstein's criticisms of Dedekind, we can remind them that he was in good company: Frege and Russell.

[22] The remark concerning Dedekind is not quite true because, as we saw, Dedekind thought that the cut was a purely arithmetical phenomenon. Unless, of course, he confused arithmetic and set theory, which seems to have been the case. There is therefore nothing terribly wrong with Wittgenstein's remarks if they are seen as a reaction against this particular misconception.

8

Strict Finitism

8.1. DUMMETT'S INTERPRETATION

The philosophical world outside Cambridge discovered the later Wittgenstein's philosophy of mathematics with the publication in 1956 of the *Remarks on the Foundations of Mathematics*, a selection from manuscripts dating from 1938 to 1944. In his 1958 survey paper, 'Eighty Years of Foundational Studies', Hao Wang promoted strict finitism, which he then called 'anthropologism', to the rank of a foundational thesis alongside finitism, intuitionism, predicativism, and Platonism, with the later Wittgenstein being enrolled as its most important representative (Wang 1958: 473–6). Within a year review articles by Paul Bernays, Michael Dummett, and Georg Kreisel were published in which Wittgenstein's philosophy was construed, for various reasons, as strict finitist (Bernays 1986; Dummett 1978a: 248–68; Kreisel 1958a). Dummett, on whose interpretation I shall concentrate in this section, wrote that 'Wittgenstein's constructivism is of a much more extreme kind than that of the intuitionists' (1978a: 180) and two pages later, after presenting what he took Wittgenstein's arguments to be, he added that 'this constructivism, more severe than any other version yet proposed, has been called "strict finitism" by Kreisel and "anthropologism" by Hao Wang' (p. 182). This has been by far the prevailing view ever since, with some exceptions.[1] But what does this claim really amount to? A useful preliminary step would be to give a brief characterization of strict finitism. Various strict finitist programmes have been devised in the past, and they do not show much homogeneity: contrary to intuitionism, there is no orthodoxy. I shall therefore limit myself to presenting, using freely the writings of

[1] To take a recent example, in the introduction to their book *Constructivism in Mathematics: An Introduction*, in which they give a brief survey of the different schools of constructivism, Anna Troelstra and Dirk van Dalen group Wittgenstein with A. S. Esenin-Volpin as representatives of strict finitism (Troelstra and van Dalen 1988: 29).

Bernays, Esenin-Volpin, Gandy, and others, a few features of the various strict finitist programmes, of interest to the debate surrounding the interpretation of Wittgenstein. (The following remarks are therefore not meant as a rigorous, exhaustive presentation.) These features are the critique of the following two assumptions:

(A1) The uniqueness up to isomorphism of the natural number series.
(A2) The principle of mathematical induction from n to Sn

and the introduction of the notion of 'feasible' number. After presenting these, I shall give a brief account of Dummett's interpretation of Wittgenstein as a strict finitist and criticize it.

The central theme of strict finitism was first formulated by Paul Bernays in 'On Platonism in Mathematics' (Bernays 1983), in the midst of a discussion on intuitionism. Bernays wanted to show that intuitionist notions contain an essential element of idealization which is absent from the finitist systems associated with Hilbert's programme, and he did so pointing out that intuitionist restrictions on acceptable existential assertions in mathematics are not as strong as they seem:

> Intuitionism makes no allowances for the possibility that, for very large numbers, the operations required by the recursive method of constructing numbers can cease to have a concrete meaning. From two integers k, l one passes to k^l; this process leads in a few steps to numbers which are far larger than any occurring in experience, e.g. $67^{257^{729}}$. Intuitionism, like ordinary mathematics, claims that this number can be represented by an Arabic numeral. Could not one press further the criticism which intuitionism makes of existential assertions and raise the question: What does it mean to claim the existence of an Arabic numeral for the foregoing number, since in practice we are not in a position to obtain it? (p. 265)

To begin with a number such as 2^{63553} is not, strictly speaking, an expression in the decimal system. It consists indeed of two decimal numbers whose position shows what operation must be performed on them to obtain the number in the decimal system, i.e. to multiply 2 by itself 63553 times, and it is usually assumed that, although it is practically impossible to carry out this task and thus to obtain the number in the decimal system which corresponds to 2^{63553}, the task is feasible in principle. Bernays introduced for the first time in this passage a distinction between possibility *in practice* and possibility *in principle*. (What is possible in practice will later be called the 'feasible'.) A number such as 2^{63553} ceases to have, in Bernays' words, 'concrete meaning', since it is not practically possible to obtain the number in the decimal system which corresponds to it. Does such a number 'exist' even if we can never write it down? While the intuitionist would be satisfied, along with the classical mathematician, with the simple notion of possibility in principle, the strict finitist asks that the distinction between possibility *in practice* and possibility *in principle* be taken seriously. Some strict finitists would actually want to reject mathematical results which depend on possibility in principle alone, while others would just like to study the distinction for its own sake.

Although Bernays did not consider the distinction as really meaningful, others did. For example, the Dutch mathematician Dirk van Dantzig, of intuitionistic leanings, contended that if the intuitionist was consistent, he would simply not call $10^{10^{10}}$ a finite number (van Dantzig 1955: 277). Well-known mathematicians had already voiced their doubts. Émile Borel, for reasons linked with his results in probability theory, expressed doubts of a strong strict finitist flavour as early as 1927: 'That we should consider a number as virtually known when its computation is theoretically possible but needs an amount of time and effort out of proportion with human possibilities seems to me a more serious question' (Borel 1927: 272). Borel gave the following example: let us take the first four decimals of π, which are 1415. We then calculate the first 1415 decimals of π, a calculation which is still possible. But suppose that we go on calculating in turn the number of decimals equivalent to 1415 numerals, etc.

> If we go on the same way a thousand times, we would define numbers whose practical calculation would not only need a multitude of human lives, but, even supposing they were known, whose writing would necessitate an amount of paper whose weight would be largely superior to that of the globe.
>
> Should we consider that the last numeral of the thousandth number so defined is calculable by us? (p. 272)

Towards the end of his life Borel came to the conclusion that 'when the finite becomes very large, it raises the same difficulties as the infinite' (Borel 1972: 979), and went on writing an almost completely forgotten book, *Les Nombres inaccessibles*, where he distinguishes at the outset 'accessible' numbers from numbers 'inaccessible' relative to decimal notation and the length of human life (Borel 1952: 2).

As I have said, Hao Wang had in 1958 promoted strict finitism to the rank of a foundational thesis alongside finitism, intuitionism, predicativism, and Platonism, giving renewed credibility to the idea. Nevertheless, as a foundational programme it remains to this day largely neglected.[2] As far as mathematical logic is concerned, the Russian A. S. Esenin-Volpin was for a long time virtually alone in trying to develop a strict finitist programme under the name of 'ultra-intuitionism' (Esenin-Volpin 1961; 1970; 1981).[3] He reasoned on lines similar to Bernays and Borel: not only does the actual infinite not exist but there is also no difference between the large finite and the potential infinite, since it is practically impossible for us to reach some large finite numbers. This is why the large finite should be understood as potential. His programme is meant to be, following Bernays' remark quoted above, a radicalization of the intuitionist critique:

> I accept the traditional intuitionistic criticism of Brouwer and go further. I ask why has such an entity as 10^{12} to belong to a natural number series? Nobody has counted up to it (10^{12} seconds constituting more than 20 000 years) and every attempt to construct the 10^{12}-th member of the sequence 0, 0', 0'', . . . requires just 10^{12} steps. But the expression '*n* steps' presupposes that *n* is a natural number i.e. a number of a natural number series. So this natural attempt to construct the number 10^{12} in a natural number series involves a vicious circle. This vicious circle is no better than that involved in the impredicative definitions of set theory: and if we have proscribed these definitions we have to proscribe the belief in the existence of a natural number 10^{12}, too. (Esenin-Volpin 1970: 4–5)

This passage, which has to do with (A2) above, reveals an important aspect of the strict finitist radicalization of the intuitionist critique of classical mathematics. Numerals are usually defined as follows: 0 is a numeral, and if *n* is a numeral, then S*n* is a numeral. Thus the numerals are

[2] Erwin Engeler expressed the received view on strict finitism when he said that it cannot reasonably be taken as a foundation of mathematics and that 'it would be more realistic to leave the theory—or a more appropriate variant of it—where it arose: in computer science, as a theory of feasible processes on a computer' (1981: 355–6).

[3] It is fair to say that Esenin-Volpin's peculiar programme, with his own methods of investigation which takes 'extreme directions' (with such strange names as 'extra-ultra-intuitionism', 'trans-ultra-intuitionism', and 'pragma-ultra-intuitionism': Esenin-Volpin (1970: 44–5)) has attracted little attention. (He introduced, however, some changes to his programme in Esenin-Volpin (1981) in order to reduce the dependence on these extreme directions.) One author who has been influenced by Esenin-Volpin's ideas is David Isles: see Isles (1981).

0, S0, SS0, SSS0, SSSS0 . . .

It is also usually assumed that with such a series one can reach any natural number, *in principle*. The truth of the induction schema $I(\phi)$ of Peano Arithmetic,

$$[\phi(0) \wedge \forall x\, (\phi(x) \rightarrow \phi(Sx))] \rightarrow \forall x\, (\phi(x))$$

relies on our recognition of this fact. Let us call this traditional natural number series **N**. In the above quotation Esenin-Volpin is asking us to consider more closely assertions such as:

$$[\phi(0) \wedge \forall x\, (\phi(x) \rightarrow \phi(Sx))] \rightarrow (\phi(10^{12}))$$

It seems that we would need a sequence of 10^{12} times the Modus Ponens, with each of them so arranged that the consequent of each (except the last) is the antecedent of the next, and that together with the antecedent of the first they imply the consequent of the last.[4] Esenin-Volpin is asking here: how could we use 10^{12} times the Modus Ponens, since in seconds 10^{12} constitutes more than 20,000 years? Obviously, nobody has counted up to 10^{12}. (If 10^{12} seems too small, one just has to pick a larger number.) As Dummett once pointed out, the common justification of the procedure of proof by induction takes the form of a repeated Modus Ponens:

F(0)
F(0) → F(S0)
F(S0) → F(SS0)
F(SS0) → F(SSS0)
. . . F(n)

Since each of these Modus Ponens is true, then F(n) is also true (Dummett 1978a: 251). For strict finitists, an explanation such as Dummett's is simply not acceptable, because it is circular: induction is needed to justify that each Modus Ponens is true (Isles 1981: 112).[5]

[4] There are of course shorter ways of reaching numbers such as 10^{12}. For example, George Boolos has shown that there is a derivation in Mates's system of natural deduction (Mates 1972), of the equivalent of SSS . . . S0 with 1000000 occurrences of S, which is less than 70 lines long (Boolos 1991: 699–700). There is no need for us to go into the details of the consequences for 'feasiblity' raised by such short derivations. For the purposes of interpreting Wittgenstein's remarks, it suffices to remain at a rather unsophisticated level.

[5] Dummett is quite aware of that fact. See Dummett (1978a: 252). This criticism is widespread: see Parikh (1971: 502–3) and Nelson (1986: 174).

These considerations led Esenin-Volpin to the introduction of the notion of 'feasible' natural numbers or *nombre réalisable ou exécutable*, i.e. numbers up to which it is still practically possible to count (Esenin-Volpin 1961: 203). The number 10^{12} could not be in this series since, according to Esenin-Volpin, it is not feasible. It acts as an *upper bound* to the series of feasible numbers.[6] Here is a definition of the feasible numbers, with F standing for 'feasible':

(α) $F(0)$	(The number 0 is feasible)
(β) $F(n) \rightarrow n<10^{12}$	(If n is feasible, then $n<10^{12}$)
(γ) $F(n) \rightarrow F(Sn)$	(If n is feasible, then Sn is also feasible)

According to this definition, each feasible natural number can be obtained—in the usual way—from 0 by going along the series:

0, S0, SS0, SSS0, SSSS0 . . .

Let us call this series **F**. It is a natural number series, being closed under the function S (successor).[7] It is not closed, however, under exponentiation.[8]

Esenin-Volpin's famous example of such a series, which he calls 'natural' or 'concrete', is that of 'the number of heartbeats in my childhood' (1961: 203). Indeed, this series is bounded above: one just has to take a sufficiently large number, such as 10^{12}, and every heart-

[6] There are various ways of establishing this arbitrary bound. (Ultimately, the size of the bound is simply irrelevant.) In 'Limitations to Mathematical Knowledge', Gandy assumed the following as an hypothesis about the size of inscriptions considered by mathematicians: 'There is an upper bound B, independent of time, to the size of inscriptions. In particular if $n > 2^B$ then the numeral for n cannot be concretely presented' (Gandy 1982: 130). He gave the following estimates in binary digits: between 10^{14} and 10^{20} for the size of inscriptions considered 'in practice', and for the limitation 'in principle' (i.e. using galaxies as computers), somewhere between 10^{50} and 10^{500} (p. 131).

[7] For more radical strict finitists, no series is closed under the operation successor. For example, Jean-Paul Van Bendegem (1987: 16) studies the consequences of limiting the activity of his finite empirical mathematician to the manipulation of symbols on a piece of paper of 600 cm^2. Assuming that the smallest sign identifiable by the mathematician is a rectangle of size 0.2×0.4 cm, a stroke would occupy a surface of 0.08 cm^2 and he would be able to put $600/0{,}08 = 7500$ sign on the paper. This is Van Bendegem's very small estimate. (Of course, a change of notation from, say, the stroke system to the decimal system will increase this limit.) The finite empirical mathematician must recognize such a bound and cannot even apply the function 'successor' to it (Van Bendegem 1987: 13–16). On the other hand, Esenin-Volpin (1970) contains an inductive construction from a number notation system closed under successor of other number notation systems closed under exponentiation, Ackermann function, and faster-growing recursive functions (pp. 38–41; see also Isles 1981: 118–19).

[8] See also Van Bendegem (1987: 42–3).

beat in my childhood was followed by another heartbeat in my childhood. In 'Wang's Paradox', Michael Dummett characterized series such as **F** as being both weakly infinite (there exists a well-ordering of it with no last member, as in (γ)) and weakly finite (for a finite ordinal n, there exists a well-ordering with no nth member, as in (β)) (Dummett 1978a: 258). The number 10^{12} is not in **F**, but it is included in the traditional natural number series **N**. It seems that we have therefore two distinct natural number series, **F** and **N**, which have different 'lengths'. To recognize that fact amounts to rejecting (A1), i.e. the uniqueness up to isomorphism of the natural number series.[9] The preceding train of thought is nicely summarized in the following quotation:

> Let us consider the series F of *feasible* numbers, i.e. of those up to which it is possible to count. The number 0 is feasible and if n is feasible then $n< 10^{12}$ and so n' also is feasible. And each feasible number can be obtained from 0 by adding '; so F forms a natural number series. But 10^{12} does not belong to F.
>
> Nevertheless the traditional natural number series containing 10^{12} is generally supposed to exist. And if we accept it there are at least two different natural number series. (Esenin-Volpin 1970: 5)

Another important aspect of Esenin-Volpin's strict finitism is the consideration of 'extreme situations', such as running out of space or time to do the calculations. These situations are those in which, by lack of space or memory, errors in inscribing, reading, or computing will arise that cannot be corrected without avoiding making another set of errors, etc. For example, a mathematician producing successive numerals using a stroke notation | || ||| |||| ||||| . . . will obviously get into an extreme situation long before a second one who uses arabic numerals and positional notation.

To describe strict finitist philosophy in one simple, basic thesis, it can be said that it is a radicalization of the intuitionist standpoint: mathematical existence is linked to practical limitations of humans to carry out operations. So it is the argument about our epistemic limitations, which was presented in section 6.3, that motivates the introduction by strict finitists of the notion of 'feasible' numbers and leads them to the rejection of assumptions such as the uniqueness up to isomorphism of the natural number series. As I take it, the interpretation

[9] See Esenin-Volpin (1970: 38–41) for a construction of equivalents of F and N. For a clarification of Esenin-Volpin's views on uniqueness, see Isles (1981).

of Wittgenstein as a strict finitist by Dummett and others[10] is based on an amalgamation of remarks about 'surveyability' of proofs, mainly contained in later writings and the strict finitist insistence on 'feasibility'. Wittgenstein's remarks on infinity, which show that he favoured an Aristotelian conception of the infinite, suggest not only that he did not agree with any insistence on practical limitations at the expense of the potential infinite but also that he was hostile to the argument about epistemic limitations, as we saw in section 6.3. This suggests that if Wittgenstein discussed at length the topic of surveyability of proofs, his interest in it could not have originated in his recognition that 'human feasibility' must play a central role in philosophy and foundations of mathematics. One must therefore raise doubts about the cogency of the interpretation of Wittgenstein as a strict finitist. There is one specific argument which could be misconstrued as being of a strict finitist nature, which appears already in the discussions with Schlick and Waismann in 1929. This argument, as we shall see in the next and last section, marks a deepening of Wittgenstein's criticisms of the traditional aims of logical foundations. But one must first be clear about what Dummett meant when he claimed that Wittgenstein was a strict finitist. Since this interpretation of Wittgenstein is largely based on his remarks on 'surveyability' in the *Remarks on the Foundations of Mathematics* and the *Lectures on the Foundations of Mathematics*, therefore on material pertaining to the later period, a fuller critical discussion of his interpretation falls outside the scope of this book.

According to Dummett, Wittgenstein held the view that 'the sense of an arithmetical predicate such as "is prime" is given not by a method that may "in principle" be used to decide its application but by the criterion we accept in practice' (Dummett 1981: 506). Since for any predicate there will always be a number too large for the practical application of any given criterion, because the computation needed would then be unsurveyable, no predicate could be said to have a determinate sense for 'all' natural numbers. Dummett took the example of the sieve of Eratosthenes. This method was conceived in order to establish tables of prime numbers: take any integer $n \geq 2$, and consider all the integers n in the interval

[10] e.g. 'Wittgenstein makes many cryptic observations . . . which become understandable if we keep in mind his preoccupation with the conception of mathematics as a feasible activity' (Wang 1958: 474).

$I_n = [n, n^2]$

If we take out of I_n all the multiples of 2, 3, 5 . . . for all the prime numbers $< n$, then all the integers of I_n surviving the sifting are prime numbers. Indeed, if n is an integer surviving the sifting and with at least two prime factors, then n would be at least as large as n^2 and it would thus lie outside I_n. If we take the first n prime numbers p_1 . . . p_n, this sieve enables us to give the list of all prime numbers up to $(p_n+2)^2$.[11] There are obvious practical limitations to this procedure. According to Dummett, Wittgenstein would not admit on the basis of these that we have only one method of deciding for all integers whether they are prime or composite, and therefore that a predicate such as 'x is prime' is not decidable for 'all' numbers. To make his point, Dummett considers the following case of a fanatic devoting his life to computing the primality of a very large number by means of the sieve of Eratosthenes, while it has already been proved prime by a more powerful method. If the fanatic's result is that the number in question is composite, our reaction, according to Dummett, would be that we would not give up our result, but rather claim that there must be an error in the fanatic's computation. This purports to show that 'we are taking the "advanced" test, and not the sieve, as the *criterion* for primality here: we use the theorem as the standard whereby we judge the computation, and not conversely. The computation is of no use to us because it is not *surveyable*' (Dummett 1978a: 180).

One must first distinguish between those cases where the sieve can and those where it cannot be applied, because it requires too many steps (considering our limitations). One can also distinguish, within the limits of possible applications of the sieve, between those that are surveyable and those that are not. Dummett's interpretation is concerned with the latter distinction, since the interesting case is that of unsurveyable but still possible computations. As Crispin Wright put it:

Suppose we are concerned with numbers to which the application of the sieve has become unsurveyably but not *impossibly* long; that is, we are in a region of integers where we keep getting different results to each other, keep

[11] The sieve by itself cannot help us to prove the infinity of prime numbers, unless we can obtain a lower positive bound (with $z \rightarrow {}^\circ$) to the number of elements surviving the sifting. This is quite an effective means of obtaining prime numbers, but there are stronger (analytic) sieves, those of Brun, Selberg, and Linnik and Renyi.

turning up errors in our own calculations, and where successive checks keep on giving different verdicts. (Wright 1980: 126)

This is clearly an 'extreme situation' in the sense given above. If the sieve of Eratosthenes is inadequate in some cases because the computations are unsurveyable, there are stronger methods to be used—there are many well-known analytic sieves. As Dummett noticed, in those cases we will base our judgements on the stronger method and its surveyable result. But for any new method adopted there will always be a number so large that the application of the new method will be unsurveyable. Therefore, according to Dummett's Wittgenstein, 'we should have no right to assert that every number is either prime or composite' (Dummett 1978a: 181). The implications of this position are quite astounding. Even Euclid's theorem on the infinity of prime numbers would be rejected. If one rejects any appeal to the 'unfeasible', there can be no proof because the proof (see section 4.1) requires the product

$$N = [\mathrm{p}_1 \times \mathrm{p}_2 \times \mathrm{p}_3 \times \ldots \times \mathrm{p}_n] + 1$$

which for a sufficiently large n may not be feasible (or surveyable). Consequently, there must be a largest prime number, since for p_n very large, one may not be able to construct the extension N.[12] This claim was also made by Crispin Wright on Wittgenstein's behalf: 'we cannot in general guarantee the *existence* of a surveyably representable $n! + 1$ for an arbitrary surveyably representable n' (Wright 1980: 138).[13] The picture of Wittgenstein just sketched presents him as committed to a position so radical that it becomes so utterly implausible that it must be resisted. Dummett's argument about the vagueness of predicates, presented towards the end of section 6.3 above, was devised to do that job.

There is a passage in the *Lectures on the Foundations of Mathematics* where Wittgenstein and Turing discuss the distinction between calculation and experiment, taking the example of multiplication, and where Wittgenstein considered a case exactly similar to

[12] This claim was made explicitly by Van Bendegem (1987: 70): 'Take once again the theorem of Euclid: there is no largest prime number. But is there, finitistically? Since all notation systems are finite, and the total of all notation systems is finite too, the answer should be: No, there is indeed a largest prime number'.

[13] Wright's formulation is slightly misleading, since in the proof of Euclid's theorem only $[\mathrm{p}_1 \times \mathrm{p}_2 \times \mathrm{p}_3 \times \ldots \times \mathrm{p}_n] + 1$ is needed, not $n! + 1$. The factorial function grows very quickly.

the one described by Wright: 'Suppose that we make enormous multiplications—numerals with a thousand digits. Suppose that after a certain point, the results people get deviate from each other. There is no way of preventing this deviation; even when we check their results, the results deviate' (*LFM*, p. 101). Wittgenstein immediately asks: 'What would be the right result? Would anyone have found it? Would there be a right result?' (p. 101). To this last question, Dummett and Wright would be inclined to think that Wittgenstein would answer: no. But this is not Wittgenstein's actual answer: 'I should say, "This has ceased to be a calculation".' (p. 101)[14] Wittgenstein is claiming here that if a calculation is unsurveyable it loses its character of being a proof and becomes an experiment, and hence cannot properly be called a calculation. So the case of a very long unsurveyable but still possible multiplication—or of a similarly complex application of the sieve—on which Dummett and Wright are resting their case would not even be recognized as a calculation by Wittgenstein. To say that when calculations with the sieve are unsurveyably long they are no longer to be considered as calculations does not amount to saying that there are cases where there is no correct application of the predicate 'prime number'. It amounts only to saying that one should not conceive of such cases as counterexamples to the proposition: every integer is either prime or composite. After all, we are not dealing with calculations here—we are missing a main ingredient of the language-game: agreement.[15]

All this may be said to be rather unconvincing, however, because Dummett could still reply that in saying that an unsurveyably long calculation is not really a calculation, Wittgenstein implies that in that case there is no pre-existing right answer. If this is indeed what Wittgenstein implies, then Dummett has a case. But in the previous passages Wittgenstein was merely recognizing a rather undeniable fact of mathematical practice. It is not denied by, say, the intuitionist. Therefore, to recognize that very large calculations lose their character of certainty does not amount to a denial that a correct answer is *in principle* obtainable; some supplementary assumption is to be made. Dummett must show that Wittgenstein effectively claims, somewhere, that in the cases under consideration, involving 'extreme

[14] There is similar discussion of unsurveyable but not impossibly long calculations in *RFM* vii, § 9.

[15] If we were within the realm of rule-governed behaviour, there would be a correct answer, so there exists a correct answer, in principle.

situations', there is no correct answer. Here, Dummett's reading turns out to rest almost entirely on the following passage, also from the 1939 lectures, where Wittgenstein went back to the same topic:

> 'For us human beings, the best thing we can arrive at, the nearest we can get, is that we always get it, or someone who had a lot of experience always got it.' As if only a God really knew.—Turing suggested this, and that is just where he and I differ. Actually there is nothing to stop us postulating that your result is right—so that in future all your children will have to copy what is written on the blackboard. And then it is right.—There is nothing there for a higher intelligence to know—except what future generations will do. We know as much as God does in mathematics. (*LFM*, pp. 103–4)[16]

This passage seems indeed to confirm Dummett's reading, since Wittgenstein claims here that 'there is nothing to stop us postulating that your result is right', and this in turn implies that according to him there is in 'extreme situations' no pre-existing right answer (Dummett 1978b: 65). I shall make two points against Dummett's reading of this passage, denying him two premisses, without which I believe that his interpretation cannot hold.

The first point is that Wittgenstein's remarks on 'following a rule' cannot provide support for Dummett's reading. According to him, when Wittgenstein said that 'nothing can stop us postulating that your result is right' and that ' in future all your children will have to copy what is written on the blackboard', he must have meant that there is indeed no right answer pre-existing our postulating one. This means that in cases involving 'extreme situations', but also by extension in all cases, the correct result of application of rule is a matter of a decision. The rule-following argument has indeed been interpreted in such a way that our rule-governed responses are understood as 'informed not by an *intuition* (of the requirements of the rule) but a kind of *decision*' (Wright 1989: 240). The connection with Kripke's interpretation of rule-following is easily seen: one just has to think here of the decision being taken by the community (Kripke 1982: ch. 3).[17]

[16] In conversation, Michael Dummett told me that when he wrote his famous review of the *Remarks on the Foundations of Mathematics*, he had in mind precisely this passage, which he knew from a copy of the Bosanquet set of notes of the 1939 lectures. It is noticeable that Dummett has never modified his view since, and that it is this same passage which he cites in Dummett (1993: 459). John Mayberry pointed out to me, also in conversation, a passage of the *Remarks on the Foundations of Mathematics* (v, § 9) which can be said to support Dummett's interpretation.

[17] The positive aspect of Wittgenstein's remarks on rule-following was rather that it is we who determine, using the right result of the application of the rule, what is to

I believe, however, that the thrust of the remarks about rule-following, as I presented them in section 6.1, is exactly to the contrary, namely that the result of applying the rule is not, at each new step, a matter of a decision. That is what Wittgenstein meant when writing that obeying the rule is something done blindly, something that 'I do not choose' (*PI*, § 219). In section 6.1 I argued that part of the background to the remarks on rule-following is Wittgenstein's rejection of his own phenomenological conception of rules, a variant of which he attributed to the intuitionists, and we saw that this was the reason why Wittgenstein described what he understood as the intuitionist explanation of rule-following as 'all bosh'. The reader should recall that, in the same passage, Wittgenstein said, 'We might as well say that we need, not an intuition at each step, but a *decision*.—Actually there is neither. You don't make a decision: you simply do a certain thing. It is a question of a certain practice' (*LFM*, p. 237). This is also said in the *Brown Book*:

> It is not an act of insight, intuition, which makes us use the rule as we do at the particular point of the series. It would be less confusing to call it an act of decision, though this too is misleading, for nothing like an act of decision must take place. (p. 143)[18]

It is quite clear also from §§ 186 or 213 of the *Philosophical Investigations* that Wittgenstein rejected any explanation of rule-following in terms of 'intuition', and that he does not believe that an explanation in terms of some kind of decision (privately made or not)

count as complying with the rule. If $25 \times 25 = 625$ and someone gets 50, then we would say that that person has not followed the rule for multiplication. Thus, his point was not that there is or that there is not a correct answer existing prior to a decision (collective or private) that $25 \times 25 = 625$, but that this equation is now being *used* as a 'paradigm for judging experience' and that, used as such, it is not a statement (true or false) which would assert some behaviouristic (or conventional) fact such as that 'anyone who has been trained in such and such a way gets, under normal circumstances, 625 as the result of multiplying 25 by 25': 'The justification of the proposition $25 \times 25 = 625$ is, naturally, that if anyone has been trained in such and such a way, then under normal circumstances he gets 625 as the result of multiplying 25 by 25. But the arithmetical proposition does not assert *that*. It is so to speak an empirical proposition hardened into a rule. It stipulates that the rule has been followed only when that is the result of the multiplication. It is thus withdrawn from being checked by experience, but now serves as a paradigm for judging experience' (*RFM* vi, § 23). For a similar reading of Wittgenstein, see Gerrard (1990: 166).

[18] Alice Ambrose also recorded Wittgenstein as saying, at about the same time (1934–5), that '[i]f any mental process is involved, it is one of decision, not of intuition' (*AWL*, p. 134). Again, the point of the rule-following argument is precisely to show that any explanation in terms of private 'mental processes' is inadequate.

fares better. Thus, it should be clear that, according to Wittgenstein, in the case of previously uneffected calculations or in the case of calculations involving 'extreme situations' the result is never something that, say, the community postulates or decides.

However, if Wittgenstein argued that each new use of the rule does not require a decision, he must have held that there is a right result. I believe that this is precisely why he would say that we are naturally inclined to think that the steps are all already taken and it is just a matter of our writing them down (*RFM* i, § 22). This is the Platonist image of the 'rules as rails' which is discussed in § 218–19 of the *Philosophical Investigations*, and in other passages such as this one: 'My question really was: "How can one keep to a rule?" And the picture that might occur to someone here is that of a short bit of hand-rail, by means of which I am to let myself be guided further than the rail reaches' (*RFM* vii, § 66). According to this image, it is not just that one is guided further by the rail into, so to speak, uncharted territory, but that the result of the previously uneffected calculation that one is about to make is already there, waiting to be discovered. In the *Remarks on the Foundations of Mathematics*, Wittgenstein wrote:

> In his fundamental law Russell seems to be saying of a proposition: 'It already follows—all I still have to do is, to infer it'. Thus Frege somewhere says that the straight line which connects any two points is really already there before we draw it; and it is the same when we say that the transitions, say in the series +2, have really already been made before we make them orally or in writing—as it were tracing them. (*RFM* i, § 21)[19]

There is thus a tension in Wittgenstein's thought between, on the one hand, the rejection of this Platonist conception of the steps of the rules being already taken and looking like the shadow of a reality waiting to be discovered and, on the other hand, the need to recognize that at each new step there is only one correct way of following the rule, only one right result in its application, this being tantamount to saying that the steps are already taken. Wittgenstein's challenge consisted of avoiding the spell of the Platonist image.[20]

[19] In this passage one can see clearly the connection (through the reference to Frege (1966: § 66) already quoted in section 3.3) between the anti-Platonism which motivates Wittgenstein's argument against numerical equivalence and his rejection of the imagery of 'rules-as-rails' underlying the phenomenological account of rule-following.

[20] At least one may agree with Juliet Floyd's remark: 'Wittgenstein has not *denied* that there is any sense in which a formula can be said to determine the steps' (Floyd 1991: 155). According to her, the gist of Wittgenstein's discussion, especially in the first sections of *RFM*, is that there not a single notion of determination (pp. 155 f.).

This image of the rule is, for Wittgenstein, rooted in the phenomenological conception of rules which was presented in section 6.1. This passage from the *Brown Book* demonstrates this clearly:

> 'But surely if one had asked me which number he should write after 1568, I should have answered "1569".'—I dare say you would, but how can you be sure of it? Your idea really is that somehow in the mysterious act of *meaning* the rule you made the transitions without really making them. You crossed all the bridges before you were there. (p. 142)

So a phenomenological account of rule-following would thus provide support to Platonism, and the remarks on rule-following might be said to undermine Platonism by showing *inter alia* that this account is inadequate. I do not see how such a limited conclusion necessarily implies the adoption of a strict finitist stance, and I fail to see that Wittgenstein's remarks are not compatible with the 'potentialist' approach to infinity which he adopted (see section 6.3 above).

The second point that I wish to make against Dummett's reading (of *LFM*, pp. 103–4, above) is that a mere rejection of the Platonist's appeal to God—who knows the answer when we cannot obtain it—is not enough to qualify Wittgenstein as a strict finitist. Intuitionists also reject this appeal to God (see the end of section 6.3 above); so when Wittgenstein said, 'We know as much as God does in mathematics', he was not out of line with them. But Dummett would be inclined to think that Wittgenstein was making a different, stronger claim than that of the intuitionist: while the latter would claim that the appeal to God in order to preserve the classical notion of truth (and avoid having recourse to intuitionistic logic) was outside the domain of decidability, Wittgenstein was supposedly making a claim within, so to speak, the domain of decidability, namely that there is nothing outside the domain of the 'feasible' which would be known by God. Therefore, whenever we reach the domain of the unsurveyable, any new application of the rule calls for a decision because there is no right answer known by God. We have just seen that such a view cannot be attributed to Wittgenstein, because applying a rule is not, for him, a matter of 'decision'.

Moreover, this view would also imply in turn that Wittgenstein wanted to draw the epistemological distinction between feasibility and possibility in principle. Again, such a distinction is explicitly rejected by him (see section 6.3 above). If I am right, Wittgenstein wanted to reject as wrong-headed the debate between the strict finitist and the Platonist around the epistemological notion of our practical limitations—not to take sides. In fact, in the passage under contention he attributes the epistemological characterization to Turing (rightly or wrongly, I do not know) and indicates clearly that he disagrees: ' "For us human beings, the best thing we can arrive at, the nearest we can get, is that we always get it, or someone who had a lot of experience always got it." As if only a God really knew.—Turing suggested this, and that is just where he and I differ' (*LFM*, p. 103). The argument that follows is admittedly badly put—by Wittgenstein or by those who wrote down the notes?—and has caused confusion. But Wittgenstein's stance is clearly stated, and it is in line with the rest of his views. Does he not say later on that 'we want to see the absurdities both of what the finitists say and of what their opponents say' (*LFM*, p. 111)?[21]

8.2. COMPLEXITY

On a more positive note, I would like to claim that Wittgenstein's insistence on the surveyability of proofs was meant to provide him with an argument against the 'reductionist' claim of logicism and its epistemological counterpart.[22] This leads us to the argument found in the notes of discussions with Schlick and Waismann which was announced earlier (*WVC*, p. 35). This argument was also alluded to in section 3.2 above: Wittgenstein was using it to argue against the view held prominently by Ramsey that mathematics is of a tautological character.

One may wish to distinguish two logicist claims. The first is that all theorems of, say, arithmetic can be deduced from purely logical axioms—that we can deduce the axioms of Peano Arithmetic from those of (second-order) logic—and, correspondingly, that arithmetical concepts are reducible to logical concepts. This reductionist claim is precisely *the* foundational claim of logicism. There seems also to be a corresponding epistemological claim which consists of the supposition that any person possessing a knowledge of logic, but is in a state of complete arithmetical innocence, can get to learn arithmetic

[21] Again, it should be clear that the position attacked by Wittgenstein in this passage (and similar ones) as 'finitism' is in fact closer to the strict finitist position discussed in this section than the kind of finitism I have already attributed to him. See section 6.3.

[22] A similar claim is made in Wright (1980: 134).

through logic alone. (Mark Steiner has focused on this claim in his book *Mathematical Knowledge* (1975), about which more later.) As we shall see, Wittgenstein's argument is an attack against both the foundational claim and its epistemological counterpart. He was by the same token raising an argument against the possibility of a reduction (in the technical sense of the term) of arithmetic to logic and defending the priority of arithmetical notions over class- or set-theoretical ones. It has already been recognized long ago, by Kreisel, that Wittgenstein's arguments in *Remarks on the Foundations of Mathematics* 'appl[y] generally to *proof theoretical reductions* or *translations*, either mapping of proofs of one kind into proofs of another, or theorems (provable formulae) of one system into theorems of another' (Kreisel 1958a: 149).[23] The gist of the argument is this: there is in fact no reduction, but a simple translation to class theory or set theory. If this is the case, then the epistemological claim has no foundation.

From a logicist standpoint of the kind Carnap tried to defend in the 1930s, arithmetical truths are nothing but logical truths systematically abbreviated, and these logical truths are in turn construed as 'analytic'. We saw in section 2.3 some of the difficulties raised by Carnap's attempt. One further problem with this approach is that most arithmetical truths would not be recognizably analytic. As A. J. Ayer pointed out a long time ago in *Language, Truth and Logic*, 'even so simple a tautology as "$91 \times 79 = 7189$" is beyond the scope of our immediate apprehension' (Ayer 1946: 86). In order to see that 7189 is synonymous with 9×79, one must calculate, and here multiplication tables are said to be 'rules for carrying out this process in arithmetic, just as the laws of logic are rules for the tautological transformation of sentences expressed in logical symbolism' (Ayer 1946: 86). We can see what would be such 'tautological transformations' by adapting, say, Quine's notion of 'numerically definite quantifier' in *Methods of Logic* (Quine 1962: 231–2). We could then give the Russellian equivalent (R) of, say, $200 + 200 = 400$ in the following *abbreviated* form:

(R) $[\,(\exists!_{200}\, x)\, Gx \wedge (\exists!_{200}\, x)\, Hx \wedge (\forall x) \neg (Gx \wedge Hx)\,] \rightarrow (\exists!_{400}\, x)$ $(Gx \vee Hx)$

It is usually assumed that there is no circularity in the fact that the numerals 200 and 400 occur in (R), because we could expand it and

[23] This is also the line followed in Frascolla (1994: 152–6).

avoid the use of the numerals by using enough variables. This is where Wittgenstein's remarks about 'surveyability' come into play. The claim is that one cannot give a totally unabbreviated form of (R), because the latter would then become unsurveyable and, as a proof of the arithmetical equation 200 + 200 = 400, it would simply lose its power to convince. Wittgenstein's argument, as recorded by Waismann in 1929, is simply that we need knowledge of the arithmetical equation to see the cogency of the Russellian tautology and not vice versa:

> The correctness of an arithmetical proposition is never expressed by a proposition's being a tautology. In the Russellian way of expressing it, the proposition 3 + 4 = 7 for example can be represented in the following manner:
>
> $(E3x)\varphi x.(E4x)\psi x.\sim(Ex)\varphi x.\psi x: \rightarrow: (E7x).\ \varphi x \psi x$
>
> Now one might think that the proof of this equation consisted in this: that the proposition written down was a tautology. But in order to be able to write down this proposition, I have to *know* that 3 + 4 = 7. The whole tautology is an application and not a proof of arithmetic. (*WVC*, p. 35)[24]

One may start counting the variables in (R), but this would simply defeat the purpose of proposing the tautological transformation as the guarantor of its truth and analyticity. Commenting in 'Process and Existence' on the increase in complexity involved in the translation of 7 + 5 = 12 along the lines of (R), Hao Wang noticed that an expanded proof of (R) would indeed require that we count the variables and this would indeed defeat the purpose of the reduction: 'We are able to see that [(R)] is a theorem of logic only because we are able to see that a corresponding arithmetic proposition is true, not the other way round' (Wang 1961: 335). So it seems that the gist of Wittgenstein's argument is that there is a circularity in Russell's attempt to found arithmetic on logic: in order to understand the logical truth, one must introduce precisely arithmetical knowledge which is meant to be proved true by the logical truth. To put it slightly differently, Wittgenstein has a point which, in Kreisel's words, 'concerns metamathematical methods used for investigating relations between two systems' (Kreisel 1958a: 150). (The argument thus works at the level of metamathematics, and it seems obvious that Wittgenstein did not recognize it as such, his various formulations thus being somewhat inadequate.) The point is simply this: it is

[24] To be honest, I have not suceeded in finding, in the 3 volumes of *PM*, any formula such as (R)!

easy to set up a language such as, say, that of second-order logic or of set theory, and to *translate* into it a formalized arithmetic (classical or intuitionist); but to speak properly of a 'reduction', the metamathematical methods used for such a translation and its study must be in some sense 'weaker' than the methods studied. The point of a reduction being that the theory to which a given theory is 'reduced' has more conceptual interest (it is 'weaker', i.e. its concepts are more readily understood, less dubious, etc.), and if one is assuming in the metamathematics necessary for the translation more than in this conceptually more interesting theory, then the purpose of the exercise is lost. Thus Wittgenstein's argument, however crudely put by him, amounts to this perfectly good point: 'the metamathematics required for a translation of arithmetic into logic requires some arithmetical concepts' (Kreisel 1958a: 150). Now, it is not to be denied that, as Kreisel points out, 'without being a reduction such a translation can be of central importance' (Kreisel 1958a: 150). But the importance is here a mathematical one, and as far as philosophy is concerned, the victim is precisely the epistemological claim of logicism: it is difficult to see how the epistemological claim that any person possessing knowledge of logic and at the same time in a state of complete arithmetical innocence can get to learn arithmetic through logic alone can survive if it is admitted that the reduction of arithmetic to logic involves arithmetic concepts.

It can be shown that this simple argument motivates Wittgenstein's controversial remarks about surveyability, and thus not the argument about epistemic limitations (only indications will be given here). For example, he argued that for proofs in *Principia Mathematica* to be epistemologically prior to their arithmetical equivalent, they must possess a 'visual' or 'geometrical' certainty (*RFM* iii, § 43), and their certainty therefore stands and falls with their geometrical cogency (*WVC*, pp. 66–7; *RFM* iii, § 16). Now, since the logical translation *increases complexity* to the extent that the tautological transformations become quickly unsurveyable, the Russellian proofs quickly lose their cogency.[25] If one looks at the difference between the Russellian

[25] It is worth noticing that Wittgenstein's argument, although it was always levelled against *PM*, applies also to the reduction to the theory of operations found in *TLP* (6.2–3, 6.241), which was developed in detail in Frascolla (1994: 8–23) and was presented in section 2.1. See e.g. the arithmetical core of the proof of the equivalent of $(2 + 3) \times (2 + 1) = 15$ in the theory of operations, i.e. $\Omega^{((2+3)\times(2+1))}{}'x = \Omega^{15}{}'x$, in Frascolla (1994: 18–19). Frascolla's conjecture is that Wittgenstein sets up in these propositions a 'theory of operations', with a view to showing that a given equation

tautology and its corresponding arithmetical equation as similar to the difference between a proof in stroke notation and its shorter version in decimal notation, Wittgenstein's argument takes the form of the following question: 'How can the proof in the stroke system prove that the proof in the decimal system is a proof?' (*RFM* ii, § 54). Wittgenstein insisted in many places that it cannot be so:

> Now let us imagine the cardinal numbers explained as 1, 1 + 1, (1 + 1) + 1, ((1 + 1) + 1) + 1, and so on. You say that the definition introducing the figures of the decimal system are a mere matter of convenience, the calculation 703000 × 40000101 could be done in that wearisome notation too. . . .
>
> Now I ask: could we also find out the truth of the proposition 7034174 + 6594321 = 13628495 by means of a proof carried out in the first notation?—Is there such a proof of this proposition?—The answer is: no. (*RFM* iii, § 3)

The reason is, in Wittgenstein's words, that one simply cannot 'command a clear view' of such a proof: 'sie ist ja nicht übersehbar' (*RFM* iii, § 3).[26] This is why Wittgenstein claimed that there is a danger in looking 'at the shortened procedure as a pale shadow of the unshortened one' (*RFM* iii, § 19). On the contrary, the shortened procedure (decimal notation) tells us the outcome of the unshortened procedure (stroke notation):

> I can find out that 100 × 100 equals 10000 by means of a 'shortened' procedure.

$t = s$ is a theorem of arithmetic if and only if one has a proof of the corresponding equation $\Omega^{t}x = \Omega^{s}{}'x$ in the theory of operations. Ultimately, the cogency of the proof in the theory of operations relies on its 'visual' aspect, and Wittgenstein's remarks about surveyability apply *mutatis mutandis*. It might be doubted, by those misguidedly fearing the spectre of psychologism, that Wittgenstein ever held the view that, as I put it, the cogency of an equation relies on its 'visual' aspect. One ought to recall, however, that Wittgenstein believed that the 'essential mathematical method is working with equations' (*TLP*, 6.2341) and that 'what is essential about equation is that it is not necessary in order to show that both expressions, which are connected by the sign of equality, have the same meaning: for this can be *perceived* from the two expressions themselves' (*TLP*, 6.232). Furthermore, Wittgenstein adds to this that 'that the propositions of mathematics can be proved means nothing else than that their correctness can be *seen* without our having to compare what they express with the facts as regards correctness' (*TLP*, 6.2321) (my emphasis in both cases). Perhaps I should also quote this remark from the 1931 lectures, regarding internal relations: 'An internal relation holds by virtue of the terms being what they are. Inference is justified by an internal relation which we see; the only justification of the transition is our looking at the two terms and seeing the internal relation between them' (*LWL*, p. 57).

[26] See also *RFM* iii, § 6.

Then why should I not regard *that* as the original proof procedure?

A shortened procedure tells me what *ought* to come out with the unshortened one. (Instead of the other way round.) (*RFM* iii, § 18)

Obviously, a proof in stroke notation of 200 + 200 = 400 is unsurveyable (*RFM* iii, § 24), and so would be the Russellian tautological transformation, which cannot therefore fulfil its role of being the 'proof' of the arithmetical equation (*RFM* iii, § 14). It is thus wrong to consider (R) as merely an abbreviation of the full-length tautology, which cannot serve as guarantor of the cogency of the shorter, surveyable proof in decimal notation: 'I want to say; if you have a proof-pattern that cannot be taken in and by a change of notation you turn it into one that can, then you are producing a proof, where there was none before' (*RFM* iii, § 2). It is important in this context to understand that Wittgenstein's argument rests not just on the fact that the change in complexity implied by the change of notation is to be taken into account, but also on the fact that the introduction of a new notation usually means more than the introduction of a method of abbreviation. Wittgenstein was pointing out against Russell that the positional notation is not dependent on the stroke notation but has a life of its own:

If there is something true about what I am trying to say, then—e.g.—calculating in the decimal notation must have its own life.—One can of course represent any decimal number in the form:

$$\frac{||||||||||||||||||||||||}{||||||||||||||||||||}$$

and hence carry out the four species of calculation in this notation. But the life of the decimal notation would have to be independent of calculating with unit strokes. (*RFM* vii, § 51)[27]

Wittgenstein's discussion of exponentiation is also instructive in this regard: according to him introducing exponentiation also means introducing new numbers, via the introduction of a new technique (*RFM* iii, § 47). It is also part of the epistemological claim that one can arrive at multiplications in arithmetic from tautological transformations without introducing new concepts. No doubt Wittgenstein's argument is to the effect that this is not possible. As he puts it: 'In this connexion, the following point constantly occurs to me: while

[27] See also *RFM* iii, § 12.

indeed a proposition 'a : b = c' can be *proved* in Russell's logic, still that logic does not teach us to construct a correct sentence of this form, i.e. does not teach us to *divide*' (*RFM* iii, § 52).

In *Mathematical Knowledge* Mark Steiner gave a rejoinder to Wittgenstein, claiming that 'anyone with training and intelligence could do decimal calculation in Russell's system' (Steiner 1975: 49). Steiner makes good use of the fact that natural numbers can be seen as either numerals or polynomials, a fact of which Wittgenstein was seemingly unaware. A numeral such as 4 denotes the successor of 3, and corresponds to SSSS0. But each numeral can be written as a polynomial in 10, in the following manner:

$$13628495 = 1\times 10^{7} + 3\times 10^{6} + 6\times 10^{5} + 2\times 10^{4} + 8\times 10^{3} + 4\times 10^{2} + 9\times 10 + 5$$

On this basis, Steiner was able to take up Wittgenstein's challenge and provide us with an abbreviated version of 7034174 + 6594321 = 13628495 which he claims is surveyable, and therefore satisfactory (Steiner 1975: 44). It suffices to assume that 0 is 0, 1 is S0, 2 is SS0 . . . ., 10 is SSSSSSSSSS0, to be able to rewrite the above equation in the following surveyable form:

$$S^{7}0\times S^{10}0^{S^{6}0} + 0\times S^{10}0^{S^{5}0} + S^{3}0\times S^{10}0^{S^{4}0} + S^{4}0\times S^{10}0^{S^{3}0} + S0\times S^{10}0^{S^{2}0} + S^{7}0\times S^{10}0^{S0} + S^{4}0 +$$
$$S^{6}0\times S^{10}0^{S^{6}0} + S^{5}0\times S^{10}0^{S^{5}0} + S^{9}0\times S^{10}0^{S^{4}0} + S^{4}0\times S^{10}0^{S^{3}0} + S^{3}0\times S^{10}0^{S^{2}0} + S^{2}0\times S^{10}0^{S0} + S0$$

$$S0\times S^{10}0^{S^{7}0} + S^{3}0\times S^{10}0^{S^{6}0} + S^{6}0\times S^{10}0^{S^{5}0} + S^{2}0\times S^{10}0^{S^{4}0} + S^{8}0\times S^{10}0^{S^{3}0} + S^{4}0\times S^{10}0^{S^{2}0} + S^{9}0\times S^{10}0^{S0} + S^{5}0$$

The obvious objection here is that one has introduced notation for addition, multiplication, and exponentiation which are not logical! In order to circumvent this difficulty, Steiner then proceeds to provide set-theoretical (second-order) definitions of addition, multiplication, and exponentiation. (There is no need to go into the details.) Recalling Wittgenstein's remark that exponentiation sets up a 'new connection between . . . the technique of counting factors and the technique of multiplying' (*RFM* iii, § 47), Steiner then points out: 'what Wittgenstein fails to appreciate is that this "new connection" can be expressed in the set theory itself' (Steiner 1975: 48). Another potential objection is that there could be, for that matter, numbers so big that they could not be written even as polynomials in 10, but

Steiner also shows how his set-theoretical notation helps us to write them down in a surveyable form. Steiner admits, however, that if we wish to get 'our calculations truly into primitive notation', instead of using convenient set-theoretical devices, 'our theorems and proofs would indeed be hopelessly complex' (Steiner 1975: 49).[28] But he argues further that unsurveyability is 'of a different color from the kind Wittgenstein claims exists in proofs proferred in *Principia*' (Steiner 1975: 49), his point being that one ought to replace 7034174 + 6594321 = 13628495 in set theory not with a first-order version, as Wittgenstein was seemingly doing, but with his own second-order version. It seems to me that Steiner has missed the point of Wittgenstein's remarks: arithmetical concepts have been introduced to provide the above version by Steiner of 7034174 + 6594321 = 13628495. That these admit in turn of a definition in a set-theoretical language shows nothing of itself. The translation of arithmetic into a set-theoretical language (or a language of second-order logic) is *not* a reduction in any sense of the word, and if this translation is not a reduction then there is no salvaging of the epistemological claim.

The most interesting aspect of Wittgenstein's remarks is that they force one to look at the complexity of proofs. Problems linked with complexity were far from central in the 1940s when Wittgenstein was writing, but the centre of gravity has shifted considerably in the past twenty years, with much of the work now done on fragments of

[28] Steiner claims that 'it may not be necessary' to perform some calculations; e.g. if one defines $(a^m)^n = a^{mn}$, then there is no need to perform calculations to show that $100^2 = 10^4$, it suffices to note that $2 \times 2 = 4$ (Steiner 1975: 48). But this procedure will not convince many, since 2^{65335} is, again, not a natural number, but an expression composed from two natural numbers and the question can still be raised: can one obtain the natural number corresponding to it (i.e. carry on 65334 multiplications)? So, if calculations have to be made, Steiner's definitions are of no help and the problem remains as before. A recent result by Edward Nelson shows that the only way to prove that the exponential of one numeral by another is equal to a numeral is to beg the question, by postulating for the formalization of the concept '*x* is equal to a numeral' some property going beyond the original concept of counting. (See Nelson (1986: 79) or, for a simplified version, Nelson (1994).) The rejection of the uniqueness of the natural number series is a consequence of this result. With a relativization scheme, one can still continue to consider numbers as being of the form: 0, S0, SS0, SSS0, SSSS0, . . . but only to a certain extent, as Nelson's result shows. This is seen by Nelson as the vindication of a purely formalist point of view for which 'the subject matter of mathematics is the expressions themselves together with the rules for manipulating them—nothing more' (Nelson 1986: 173). For him, 'the invention of positional notation was the creation of a new kind of number' (p. 173). This is a vindication of sorts of Wittgenstein's views.

Peano Arithmetic, on subrecursion, and on links with computational complexity.[29] So Wittgenstein put his finger on an important aspect of proofs which was largely neglected in the proof theory of his time. The lesson to be learnt here is that Wittgenstein's remarks on surveyability should not be seen as an unsuccessful attempt at defending a particularly radical philosophy of mathematics (strict finitism) by developing arguments based on the notion of our epistemic limitations, but rather as remarks from a philosopher interested in discovering new conceptual aspects, where further research will lead to fruitful results. Admittedly, Wittgenstein's remarks are very unsophisticated; in Kreisel's words they are 'too banal to be of use' (1976: 170). There are indeed no hints about which concepts should be used in analysis of proofs, i.e. to obtain a measure of complexity.[30] These remarks may indeed be of no use to specialists, but to philosophers who wish to take stock of the current trends just mentioned they might open up new avenues. But, as with other issues tackled in this book, I do not claim to have done more here than spadework.

[29] For a useful introduction, see Macintyre (1987).

[30] Such a measure of complexity (nesting of inferences) was e.g. proposed in Statman (1974) and defended in Kreisel (1976; 1977).

REFERENCES

Where an English translation is available I have used it; otherwise translations are mine (except when specifically indicated in a footnote). References to Aristotle's *Physics* in Chapter 6 follow the usual convention, and the translation I use is that of R. Waterfield (Oxford: Oxford University Press, 1996).

ABERTH, O. (1970), 'Computable Analysis and Differential Equations', in J. Myhill, A. Kino, and R. E. Vesley (eds.), *Intuitionism and Proof Theory*. Amsterdam: North-Holland, 47–52.

—— (1980), *Computable Analysis*. New York: McGraw Hill.

ABHYANKAR, S. S. (1976), 'Historical Ramblings in Algebraic Geometry and Related Algebra', *American Mathematical Monthly* 83: 409–48.

AMBROSE, A. (1935), 'Finitism in Mathematics', *Mind* 44: 186–203, 317–40.

ANDREWS, P. B. (1965), *A Transfinite Type Theory with Type Variables*. Amsterdam: North-Holland.

ANSCOMBE, G. E M. (1971), *An Introduction to Wittgenstein's Tractatus*, 4th edn. London: Hutchinson.

AUSTIN, J. (1980), 'Wittgenstein's Solutions to the Color Exclusion Problem', *Philosophy and Phenomenological Research* 41: 142–9.

AYER, A. J. (1946), *Language, Truth and Logic*, 2nd edn. London: Gollancz.

BAKER, A. (1975), *Transcendental Number Theory*. Cambridge: Cambridge University Press.

BARENDREGT, H. P. (1984), *The Lambda Calculus: Its Syntax and Semantics*, 2nd edn. Amsterdam: North-Holland.

BEESON, M. J. (1985), *Foundations of Constructive Mathematics: Metamathematical Studies*. Berlin: Springer.

—— (1993), 'Constructivity in the Nineties', in J. Czermak (ed.), *Philosophy of Mathematics: Proceedings of the 15th International Wittgenstein Symposium, Kirchberg-am-Wechsel 1992*, i. Vienna: Hölder-Pichler-Tempsky, 127–44.

BENACERRAF, P., and PUTNAM, H. (eds.) (1983), *Philosophy of Mathematics: Selected Readings*, 2nd edn. Cambridge: Cambridge University Press.

BERNAYS, P. (1983), 'On Platonism in Mathematics', in Benacerraf and Putnam (1983: 258–71).

—— (1986), 'Comments on Ludwig Wittgenstein's *Remarks on the Foundations of Mathematics*', in Shanker (1986: 165–82).

BISHOP, E. (1967), *Foundations of Constructive Analysis*, New York: McGraw-Hill.

BISHOP, E. (1975), 'The Crisis in Contemporary Mathematics', *Historia Mathematica* 2: 507–17.

—— and BRIDGES, D. S. (1985), *Constructive Analysis*. Berlin: Springer.

BLACK, M. (1964), *A Companion to Wittgenstein's 'Tractatus'*. Cambridge: Cambridge University Press.

BOLTZMANN, L. (1974), *Theoretical Physics and Philosophical Problems: Selected Writings*. Dordrecht: Reidel.

BOOLOS, G. (1987), 'The Consistency of Frege's *Foundations of Arithmetic*', in J. J. Thompson (ed.), *On Being and Saying: Essays for Richard Cartwright*. Cambridge, Mass.: MIT Press, 3–20.

—— (1988), 'Saving Frege from Contradiction', *Proceedings of the Aristotelian Society* 87: 137–51.

—— (1990), 'The Standard of Equality of Numbers', in Boolos (ed.), *Meaning and Method: Essays in Honor of Hilary Putnam*. Cambridge: Cambridge University Press, 261–77.

—— (1991), 'Zooming Down the Slippery Slope', *Noûs* 25: 695–706.

BOREL, E. (1905), 'Quelques remarques sur les principes de la théorie des ensembles', *Mathematische Annalen* 60: 194–5.

—— (1914), *Leçons sur la théorie des fonctions*, 2nd edn. Paris: Gauthier-Villars.

—— (1927), 'A propos de la récente discussion entre M. R. Wavre et M. P. Lévy', *Revue de Métaphysique et de Morale* 34: 271–6.

—— (1952), *Les nombres inaccessibles*. Paris: Gauthier-Villars.

—— (1972), *Oeuvres*, ii. Paris: CNRS.

BOUVERESSE, J. (1987), *La force de la règle: Wittgenstein et l'invention de la nécessité*. Paris: Éditions de Minuit.

—— (1992), 'Weyl, Wittgenstein et le problème du continu', in J.-M. Salanskis and H. Sinaceur (eds.), *Le Labyrinthe du continu*. Berlin: Springer, 213–29.

BRAITHWAITE, R. B. (1931), 'Editor's Introduction', in F. P. Ramsey, *The Foundations of Mathematics and Other Essays*, ed. R. Braithwaite. London: Kegan Paul, Trench Trubner, pp. ix–xiv.

BRIDGES, D. S. (1994), 'A Constructive Look at the Real Number Line', in P. Ehrlich (ed.), *Real Numbers, Generalizations of the Reals, and Theories of Continua*. Dordrecht: Kluwer, 29–92.

—— and RICHMAN, F. (1987), *Varieties of Constructive Mathematics*. Cambridge: Cambridge University Press.

BROUWER, L. E. J., (1967a), 'On the Significance of the Principle of Excluded Middle in Mathematics, Especially in Function Theory', in van Heijenoort (1967: 335–45).

—— (1967b), 'Intuitionistic Reflections on Formalism', in van Heijenoort (1967: 490–2).

—— (1975), *Collected Works*, i, ed. A. Heyting. Amsterdam: North-Holland.

BURGESS, J. P. (1993), 'Hintikka and Sandu versus Frege in re Arbitrary Functions', *Philosophia Mathematica* (ser. 3) 1: 50–65.

—— (1995), 'Frege and Arbitrary Functions', in Demopoulos (1995: 89–107).

Campbell, N. R. (1919), *Physics: The Elements*. Cambridge: Cambridge University Press, repr. 1957, *Foundations of Science*. New York: Dover.

Cantor, G. (1932), *Gesammelte Abhandlungen*, ed. E. Zermelo. Hildesheim: Georg Olms.

Carnap, R. (1931), 'Die physikalische Sprache als Universalsprache der Wissenschaft', *Erkenntnis* 2: 432–65.

—— (1937), *The Logical Syntax of Language*. London: Routledge & Kegan Paul.

—— (1967), *The Logical Structure of the World*. Berkeley, Calif: University of California Press.

—— (1983), 'The Logicist Foundations of Mathematics', in Benacerraf and Putnam (1983: 41–52).

Carruthers, P. (1990), *The Metaphysics of Tractatus*. Cambridge: Cambridge University Press.

Church, A. (1932), 'A Set of Postulates for the Foundation of Logic', *Annals of Mathematics* 33: 346–66 and 34: 839–64.

Coffa, A. (1982), 'Kant, Bolzano and the Emergence of Logicism', *Journal of Philosophy* 74: 679–89.

—— (1991), *The Semantic Tradition from Kant to Carnap: To the Vienna Station*. Cambridge: Cambridge University Press.

Cook Wilson, J. (1926), *Statement and Inference*. 2 vols., Oxford: Oxford University Press.

Crane, H. D., and Piantanida, T. P. (1983), 'On Seeing Reddish Green and Yellowish Blue', *Science* 221: 1078–9.

Curry, H. B. (1941), 'A Formalisation of Recursive Arithmetic', *American Journal of Mathematics* 63: 263–82.

Dauben, J. (1979), *Georg Cantor: His Mathematics and Philosophy of the Infinite*. Cambridge, Mass.: Harvard University Press.

Davenport, H. (1982), *The Higher Arithmetic*, 5th edn. Cambridge: Cambridge University Press.

Dedekind, R. (1932), *Gesammelte mathematische Werke*, ed. R. Fricke, E. Noether, and O. Öystein. 3 vols., Brunswick: Vieweg.

—— (1963a), 'Continuity and Irrational Numbers', in *Essays on the Theory of Numbers*. New York: Dover, 1–27.

—— (1963b), 'The Nature and Meaning of Numbers', in *Essays on the Theory of Numbers*. New York: Dover, 29–115.

—— (1967), 'Letter to Keferstein', in van Heijenoort (1967: 98–103).

—— and Weber, H. (1882), 'Theorie der algebraischen Funktionen einer Verändlichen', in Dedekind (1932: i. 238–349).

Demopoulos, W. (ed.) (1995), *Frege's Philosophy of Mathematics*. Cambridge, Mass: Harvard University Press.

—— and Bell, J. L. (1993), 'Frege's Theory of Concepts and Objects and the

Interpretation of Second-Order Logic', *Philosophia Mathematica* (ser. 3) 1: 139–56.

DIAMOND, C. (1991), *The Realistic Spirit: Wittgenstein, Philosophy and the Mind*. Cambridge, Mass.: MIT Press.

DICKMANN, M. A. (1992), 'Le Continu au premier ordre', in J.-M. Salanskis and H. Sinaceur (eds.), *Le Labyrinthe du continu*. Berlin: Springer, 117–33.

DIRICHLET, P. G. L. (1969), 'Über die Darstellung ganz willkürlicher Funktionen durch Sinus und Cosinusreihen', in *Werke*, ed. L. Kronecker. New York: Chelsea, i. 135–60.

DUMMETT, M. A. E. (1977), *Elements of Intuitionism*. Oxford: Oxford University Press.

—— (1978a), *Truth and Other Enigmas*. London: Duckworth.

—— (1978b), 'Reckonings. Wittgenstein on Mathematics', *Encounter* 50: 63–8.

—— (1981), *Frege: Philosophy of Language*, 2nd edn. London: Duckworth.

—— (1990), 'The Source of the Concept of Truth', in G. Boolos (ed.), *Meaning and Method: Essays in Honor of Hilary Putnam*. Cambridge: Cambridge University Press, 1–15.

—— (1991), *Frege: Philosophy of Mathematics*. London: Duckworth.

—— (1993), 'Wittgenstein on Necessity: Some Reflections', in *The Seas of Language*. Oxford: Oxford University Press, 448–61.

DREBEN, B., and FLOYD, J. (1991), 'Tautology: How Not to Use a Word', *Synthese* 87: 23–49.

EDWARDS, H. M. (1980), 'The Genesis of Ideal Theory', *Archives for History of Exact Sciences* 23: 322–78.

—— (1987), 'An Appreciation of Kronecker', *Mathematical Intelligencer* 9(1): 28–35.

—— (1989), 'Kronecker's Views on the Foundations of Mathematics', in D. E. Rowe and J. McCleary (eds.), *The History of Modern Mathematics*, i: *Ideas and Their Reception*. London: Academic Press, 67–77.

—— (1990), *Divisor Theory*, Boston: Birkhäuser.

—— (1992a), 'Mathematical Ideas, Ideals, and Ideology', *Mathematical Intelligencer* 14(2): 6–19.

—— (1992b), 'Kronecker's Arithmetical Theory of Algebraic Quantities', *Jahresbericht der deutsches Mathematiker-Vereinigung* 94: 130–9.

EGIDI, R. (1991), 'Ramsey and Wittgenstein on Scientific Theories', *Theoria* 57: 196–210.

ENDERTON, H. B. (1977), *Elements of Set Theory*. New York: Academic Press.

ENGELER, E. (1981), 'An Algorithmic Model of Strict Finitism', in B. Dömölki and T. Gergely (eds.), *Mathematical Logic in Computer Science*. Amsterdam: North-Holland, 345–57.

ESENIN-VOLPIN, A. S. (1961), 'Le Programme ultra-intuitionniste des fondements des mathématiques', in *Infinitistic Methods: Proceedings of the Symposium on the Foundations of Mathematics, September 1959, Warsaw*. Warsaw: PWN Warszawa, 201–23.

—— (1970), 'The Ultra-Intuitionistic Criticism and the Anti-Traditional Program for the Foundations of Mathematics' in J. Myhill, A. Kino, and R. E. Vesley (eds.), *Intuitionism and Proof Theory*. Amsterdam: North-Holland, 3–45.

—— (1981), 'About Infinity, Finiteness and Finitization', in F. Richman (ed.), *Constructive Mathematics: Proceedings of the New Mexico State Conference*. Berlin: Springer, 274–313.

FALTINGS, G. (1983), 'Endlichkeitssätze für abelsche Varietäten über Zahlkörpern', *Inventiones Mathematicae* 73: 349–66.

FEFERMAN, S. (1964), 'Systems of Predicative Analysis', *Journal of Symbolic Logic* 29: 1–30.

—— (1968), 'Systems of Predicative Analysis II: Representation of Ordinals', *Journal of Symbolic Logic* 33: 193–220.

—— (1984), 'Between Constructive and Classical Mathematics', in M. Richter, E. Börger, W. Oberschelp, B. Schinzel, and W. Thomas (eds.), *Computation and Proof Theory*. Berlin: Springer, 143–62.

—— (1987), 'Infinity in Mathematics: Is Cantor Necessary?', in G. Toraldo di Francia (ed.), *L'infinito nella scienza*/Infinity in Science. Rome: Enciclopedia Italiana, 151–209.

—— (1988), 'Weyl Vindicated: *Das Kontinuum* 70 Years Later', in *Atti del Congresso Temi e prospettive della logica e della filosofia della scienza contemporanea, Cesena, 7–10 gennaio 1987*. Bologna: CLUEB, 61–93.

—— (1993), 'What Rests on What? The Proof-Theoretic Analysis of Mathematics', in J. Czermak (ed.), *Philosophy of Mathematics: Proceedings of the 15th International Wittgenstein-Symposium, Kirchberg-am-Wechsel 1992*, i. Vienna: Hölder-Pichler-Tempsky, 147–71.

FLOYD, J. (1991), 'Wittgenstein on 2,2,2, . . . : On the Opening of *Remarks on the Foundations of Mathematics*', *Synthese* 87: 143–80.

—— (1994), 'The Continuing Significance of Moore's Objection to Wittgenstein's Discussions of Mathematics, 1930–33 and Its Bearing on *On Certainty*', in Hintikka and Puhl (1994: 163–9).

—— (1995), 'On Saying What You Really Want to Say: Wittgenstein, Gödel, and the Trisection of the Angle', in Hintikka (1995b: 373–425).

FOGELIN, R. (1968), 'Wittgenstein and Intuitionism', *American Philosophical Quarterly* 5: 267–74.

—— (1976), *Wittgenstein*. London: Routledge & Kegan Paul.

—— (1983), 'Wittgenstein on Identity', *Synthese* 56: 141–54.

FRASCOLLA, P. (1980a), 'Il concetto di infinito nella filosofia della mathematica di Wittgenstein', *Annali della Scuola Normale Superiore di Pisa: Classe di lettere e filosofia* 10: 639–74.

—— (1980b), 'The Constructivist Model in Wittgenstein's Philosophy of Mathematics', in Shanker (1986: 242–9).

—— (1984), 'Wittgenstein e il Principio del Terzo Escluso', *Rivista di Filosofia* 75: 293–309.

FRASCOLLA, P. (1994), *Wittgenstein's Philosophy of Mathematics*. London: Routledge.

FREGE, G. (1966), *Grundgesetze der Arithmetik*. Hildesheim: Georg Olms.

—— (1967), *Begriffsschrift, a Formula Language, Modeled upon that of Arithmetic, for Pure Thought*, in van Heijenoort (1967: 5–82).

—— (1980a), *The Foundations of Arithmetic*, 2nd edn. Oxford: Blackwell.

—— (1980b), 'Frege Against the Formalists: *Grundgesetze der Arithmetik*, Vol. ii, Sections 86–137', in M. Black and P. Geach (eds.), *Translations from the Philosophical Writings of Gottlob Frege*, 3rd edn. Oxford: Blackwell, 162–213.

—— (1984), *Collected Papers on Mathematics, Logic, and Philosophy*. Oxford: Blackwell.

GANDY, R. O. (1982), 'Limitations to Mathematical Knowledge', in D. van Dalen, D. Lascar, and T. J. Smiley (eds.), *Logic Colloquium '80*. Amsterdam: North-Holland, 129–46.

GARGANI, A. (1984), 'Wittgenstein on Intentional Acts', in R. Haller (ed.), *Aesthetics: Proceedings of the 8th International Wittgenstein Symposium*. Vienna: Hölder-Pichler-Tempski, i. 211–17.

GARLING, D. J. H. (1986), *A Course in Galois Theory*. Cambridge: Cambridge University Press.

GAUTHIER, Y. (1985), 'A Theory of Local Negation: The Model and Some Applications', *Archiv für Mathematische Logik und Grundlagenforschung* 25: 127–43.

—— (1991), *De la logique interne*. Paris: Vrin.

GEORGE, A. (1987), 'The Imprecision of Impredicativity', *Mind* 96: 514–18.

GERRARD, S. (1987), 'Wittgenstein in Transition: The Philosophy of Mathematics'. Dissertation, University of Chicago.

—— (1990), 'Is Wittgenstein a Relativist?', in Haller and Brandl (1990: ii. 163–73).

—— (1991), 'Wittgenstein's Philosophies of Mathematics', *Synthese* 87: 125–42.

GÖDEL, K. (1940), *The Consistency of the Continuum Hypothesis*. Princeton, NJ: Princeton University Press.

—— (1967), 'On Formally Undecidable Propositions of *Principia Mathematica* and Related Systems I', in van Heijenoort (1967: 596–616).

—— (1980), 'On a Hitherto Unexploited Extension of the Finitary Standpoint', *Journal of Philosophical Logic* 9: 133–42.

—— (1983), 'Russell's Mathematical Logic', in Benacerraf and Putnam (1983: 447–69).

—— (1986), 'Review of R. Carnap: The Logicist Foundations of Mathematics', in *Collected Works*, i, ed. S. Feferman et al. Oxford: Clarendon Press, 243–45.

GOLDFARB, W. (1979), 'Logic in the Twenties: The Nature of the Quantifier', *Journal of Symbolic Logic* 44: 351–68.

—— (1988), 'Poincaré and the Logicists', in W. Aspray and P. Kitcher (eds.), *Minnesota Studies in the Philosophy of Science*, xi: *History and Philosophy of Modern Mathematics*. Minneapolis: University of Minnesota Press, 61–81.

GOODSTEIN, R. L. (1945), 'Function Theory in an Axiom-Free Equation Calculus', *Proceedings of the London Mathematical Society* 48: 401–34.

—— (1951), *Constructive Formalism*. Leicester: Leicester University Press.

—— (1957), *Recursive Number Theory*. Amsterdam: North-Holland.

—— (1961), *Recursive Analysis*. Amsterdam: North-Holland.

—— (1965), 'Mathematical Systems', in *Essays in the Philosophy of Mathematics*. Leicester: Leicester University Press, 79–96.

—— (1969), 'A Constructive Form of the Second Gauss Proof of the Fundamental Theorem of Algebra', in B. Dejon and P. Henrici (eds.), *Constructive Aspects of the Fundamental Theorem of Algebra*. London: Wiley, 69–76.

—— (1971), *Development of Mathematical Logic*. London: Logos Press.

—— (1972), 'Wittgenstein's Philosophy of Mathematics', in A. Ambrose and M. Lazerowitz (eds.), *Ludwig Wittgenstein: Philosophy and Language*. London: Allen, and Unwin, 271–86.

GRANGER, G.-G. (1990), '*Was in Könisberg zu sagen wäre . . .* : Wittgenstein et les mathématiques', in *Invitation à la lecture de Wittgenstein*. Aix-en-Provence: Alinéa.

GRIFFIN, J. (1964), *Wittgenstein's Logical Atomism*. Oxford: Oxford University Press.

HACKER, P. M. S. (1972), *Insight and Illusion. Wittgenstein on Philosophy and the Metaphysics of Experience*. Oxford: Clarendon Press.

—— (1986), *Insight and Illusion. Themes in the Philosophy of Wittgenstein*, rev. edn. Oxford: Clarendon Press.

HAHN, H. (1980), *Empiricism, Logic, and Mathematics: Philosophical Papers*, ed. B. McGuinness. Dordrecht: Reidel.

HALLER, R., and BRANDL, J. (eds.) (1990), *Wittgenstein: Towards a Re-Evaluation. Proceedings of the 14th International Wittgenstein Symposium*. Vienna: Hölder-Pichler-Tempsky.

HALLETT, M. (1984), *Cantorian Set Theory and the Limitation of Size*. Oxford: Oxford University Press.

HARDIN, C. L. (1988), *Color for Philosophers: Unweaving the Rainbow*. Indianapolis: Hackett.

HARDY, G. H. (1929), 'Mathematical Proof', *Mind* 38, 1–25.

HECK, R. G. JR., and STANLEY, J. (1993), 'Reply to Hintikka and Sandu: Frege and Second-Order Logic', *Journal of Philosophy* 90: 416–24.

HEINZMANN, G. (1985), *Entre intuition et analyse: Poincaré et le concept de prédicativité*. Paris: Librairie Scientifique et Technique Albert Blanchard.

HENKIN, L. (1950), 'Completeness in the Theory of Types', *Journal of Symbolic Logic* 15: 81–91.

HERTZ, H. (1899), *Principles of Mechanics*. London: Macmillan.

HEYTING, A. (1958), 'Intuitionism in Mathematics', in R. Klibansky (ed.), *La Philosophie au milieu du siècle*. Florence: La Nuova Italia Editrice, 101–15.

—— (1983), 'The Intuitionist Foundations of Mathematics', in Benacerraf and Putnam (1983: 52–61).

HILBERT, D. (1935), *Gesammelte Abhandlungen*, iii. Berlin: Springer.

—— (1967a), 'On the Foundations of Logic and Arithmetic', in van Heijenoort (1967: 129–38).

—— (1967b), 'On the Infinite', in van Heijenoort (1967: 367–92).

—— and BERNAYS, P. (1934), *Grundlagen der Mathematik*, i. Berlin: Springer.

HINDLEY, J. R, and SELDIN, J. P. (1986), *Introduction to Combinators and λ-Calculus*. Cambridge: Cambridge University Press.

HINTIKKA, J. (1956), 'Identity, Variables and Impredicative Definitions', *Journal of Symbolic Logic* 21: 225–45.

—— (1989), 'Rules, Games and Experiences: Wittgenstein's Discussion of Rule-Following in the Light of his Development', *Revue Internationale de Philosophie* 43: 279–97.

—— (1990), 'Wittgenstein as a Philosopher of Immediate Experience', in Haller and Brandl (1990: i. 155–67).

—— (1992), ' "*Die Wende der Philosophie*": Wittgenstein's New Logic of 1928', in S. Teghrarian and A. Serafini (eds.), *Wittgenstein and Contemporary Philosophy*. Wakefield, NH: Longwood.

—— (1993), 'Ludwig's Apple Tree: On the Philosophical Relations between Wittgenstein and the Vienna Circle', in F. Stadler (ed.), *Scientific Philosophy: Origins and Developments*. Dordrecht: Kluwer, 27–46.

—— (1994), 'An Anatomy of Wittgenstein's Picture Theory', in C. C. Gould and R. S. Cohen (eds.), *Artifacts, Representations and Social Practice*. Amsterdam: Kluwer, 223–56.

—— (1995a), 'Standard vs. Nonstandard Distinction: A Watershed in the Foundations of Mathematics', in Hintikka (1995b: 21–44).

—— (ed.) (1995b), *From Dedekind to Gödel*. Dordrecht: Kluwer.

HINTIKKA, M. B., and HINTIKKA, J. (1986), *Investigating Wittgenstein*. Oxford: Blackwell.

—— and PUHL, K. (eds.) (1994), *The British Tradition in 20th Century Philosophy: Papers of the 17th International Wittgenstein Symposium 1994*. Kirchberg-am-Wechsel: Austrian Ludwig Wittgenstein Society.

—— —— (eds.) (1995), *The British Tradition in 20th Century Philosophy: Proceedings of the 17th International Wittgenstein Symposium*. Vienna: Hölder-Pichler-Tempsky.

—— and SANDU, G. (1992), 'The Skeleton in Frege's Cupboard: The Standard versus Nonstandard Distinction', *Journal of Philosophy* 89: 290–315.

ISAACSON, D. (1987), 'Arithmetical Truth and Higher-Order Concepts', in Paris Logic Group (eds.), *Logic Colloquium '85*. Amsterdam: North-Holland, 147–69.

ISAACSON, D. (1991), 'Some Considerations on Arithmetical Truth and the ω-Rule', in M. Detlefsen (ed.), *On the Nature of Proof: Investigations in the Philosophy of Mathematics*. London: Routledge, 94–137.

ISLES, D. (1981), 'Remarks on the Notion of Standard Non-Isomorphic Natural Number Series', in F. Richman (ed.), *Constructive Mathematics: Proceedings of the New Mexico State Conference*. Berlin: Springer, 111–34.

JAMES, W. (1983), *Principles of Psychology*. Cambridge, Mass.: Harvard University Press.

JOHNSON, W. E. (1922), *Logic*, pt. ii: *Demonstrative Inference: Deductive and Inductive*. Cambridge: Cambridge University Press.

JOURDAIN, P. E. B. (1905), 'On the General Theory of Functions', *Journal für reine und angewandte Mathematik* 128: 169–210.

KANAMORI, A. 'The Emergence of Descriptive Set Theory', in Hintikka (1995b: 241–62).

KARP, C. R. (1964), *Languages with Expressions of Infinite Length*. Amsterdam: North-Holland.

KAUFMANN, F. (1930), 'The Infinite in Mathematics and Its Elimination', in B. McGuinness (ed.), *The Infinite in Mathematics*. Dordrecht: Reidel, 1–164.

KENNY, A. (1973), *Wittgenstein*. Penguin: New York.

KOLMOGOROV, N. A. (1932), 'Zur Deutung der Intuitionistischen Logik', *Mathematische Zeitschrift* 35: 58–65.

KREISEL, G. (1950), 'Note on Arithmetic Models for Consistent Formulae of the Predicate Calculus', *Fundamenta Mathematicae* 37: 265–85.

—— (1956), 'Some Uses of Metamathematics', *British Journal for the Philosophy of Science* 7: 161–73.

—— (1958a), 'Wittgenstein's Remarks on the Foundations of Mathematics', *British Journal for the Philosophy of Science* 9: 135–58.

—— (1958b), 'Ordinal Logics and the Characterization of Informal Concepts of Proofs', in J. A. Todd (ed.), *Proceedings of the International Congress of Mathematicians*. Cambridge: Cambridge University Press, 289–99.

—— (1958c), 'A Remark on Free Choice Sequences and the Topological Completeness of Proofs', *Journal of Symbolic Logic* 273: 369–88.

—— (1960), 'La Prédicativité', *Bulletin de la Société Mathématique de France* 88: 371–91.

—— (1968), 'Lawless Sequences of Natural Numbers', *Compositio Mathematica* 20: 222–48.

—— (1970), 'Principles of Proof and Ordinals Implicit in Given Concepts', in J. Myhill, A. Kino, and R. E. Vesley (eds.), *Intuitionism and Proof Theory*. Amsterdam: North-Holland, 489–516.

—— (1976), 'Der unheilvolle Einbruch der Logik in die Mathematik', *Acta Philosophica Fennica* 28: 166–87.

—— (1977), 'On the Kind of Data Needed for a Theory of Proofs', in R. Gandy and M. Hyland (eds.), *Logic Colloquium 76*. Amsterdam: North-Holland, 111–28.

KREISEL, G. (1984), 'Frege's Foundations and Intuitionistic Logic', *The Monist* 67, 72–91.

—— (1986), 'Critical Notice: *Lectures on the Foundations of Mathematics*', in Shanker (1986: 98–110).

—— and MACINTYRE, A. J. (1982), 'Constructive Logic versus Algebraization I', in Troelstra and van Dalen (1982: 217–60).

—— and TAKEUTI, G. (1974), 'Formally Self-Referential Propositions for Cut Free Classical Analysis and Related Systems', *Dissertationes Mathematicae* 118: 1–50.

KRIPKE, S. (1982), *Wittgenstein on Rules and Private Language: An Elementary Exposition*. Cambridge, Mass.: Harvard University Press.

KRONECKER, L. (1899), *Werke*, ed. K. Hensel. 5 vols., Leipzig: Teubner.

—— (1901), *Vorlesungen über Zahlentheorie*, ed. K. Hensel. Leipzig: Teubner.

L'ABBÉ, M. (1953), 'Systems of Transfinite Types Involving λ-Conversion', *Journal of Symbolic Logic* 18: 209–24.

LAWVERE, W. (1964), 'An Elementary Theory of the Category of Sets', *Proceedings of the National Academy of Science* 52: 1506–11.

LEBLANC, H. (1975), 'That *Principia Mathematica, First Edition*, Has a Predicative Interpretation after all', *Journal of Philosophical Logic* 4: 67–70.

LEWY, C. (1967), 'A Note on the Text of the *Tractatus*', *Mind* 76: 416–23.

LORENZEN, P. (1955), *Einführung in die operative Logik und Mathematik*. Berlin: Springer.

MCGUINNESS, B. (1971), 'Philosophy of Science in the *Tractatus*', in *Wittgenstein et le problème d'une philosophie de la science*. Paris: CNRS, 9–18.

MACINTYRE, A. J. (1986), 'Twenty Years of p-adic Model Theory', in J. B. Paris, A. J. Wilkie, and G. M. Wilmers (eds.), *Logic Colloquium '84*. Amsterdam: North-Holland, 121–53.

—— (1987), 'The Strength of Weak Systems', in P. Weingartner and G. Schurz (eds.), *Logic, Philosophy of Science and Epistemology: Proceedings of the 11th International Wittgenstein Symposium*. Vienna: Hölder-Pichler-Tempsky, 43–59.

MAJER, U. (1988), 'Zu einer bemerkenswerten Differenz zwischen Brouwer und Weyl', in W. Deppert, K. Hübner, A. Oberschelp, and V. Weidemann (eds.), *Exact Sciences and Their Philosophical Foundations/Exakte Wissenschaften und ihre philosophische Grundlegung: Vorträge des Internationalen Hermann-Weyl-Kongresses, Kiel 1985*. Frankfurt a. M.: Peter Lang, 543–52.

—— (1989), 'Ramsey's Conception of Theories: An Intuitionistic Approach', *History of Philosophy Quarterly* 6: 233–58.

—— (1991), 'Ramsey's Theory of Truth and the Truth of Theories: A. Synthesis of Pragmatism and Intuitionism in Ramsey's Last Philosophy', *Theoria* 57: 162–95.

MALCOLM, N. (1984), *Ludwig Wittgenstein: A. Memoir*, 2nd edn. Oxford: Oxford University Press.

MARION, M. (1993), 'Wittgenstein and the Dark Cellar of Platonism', in K. Puhl (ed.), *Wittgenstein's Philosophy of Mathematics: Proceedings of the 15th International Wittgenstein-Symposium, Kirchberg-am-Wechsel 1992*, ii. Vienna: Hölder-Pichler-Tempsky, 110–18.

—— (1994), 'Cora Diamond on Wittgenstein and Ramsey: A Realistic Interpretation?', in Hintikka and Puhl (1994: 291–8).

—— (1995a), 'Kronecker's "Safe Haven of Real Mathematics" ', in M. Marion and R. S. Cohen (eds.), *Quebec Studies in the Philosophy of Science*, i. Dordrecht: Kluwer, 187–213.

—— (1995b), 'Wittgenstein and Ramsey on Identity', in Hintikka (1995b: 343–71).

—— (1995c), 'Wittgenstein and Finitism', *Synthese* 105: 141–76.

MATES, B. (1972), *Elementary Logic*, 2nd edn. Oxford: Oxford University Press.

MENGER, K. (1994), *Reminiscences of the Vienna Circle and the Mathematical Colloquium*, ed. L. Golland, B. McGuinness, and A. Sklar. Dordrecht: Kluwer.

MESCHKOWSKI, H. (1967), *Probleme des Unendlichen: Werk und Leben Georg Cantors*. Brunswick: Vieweg.

MOLK, J. (1909), 'Principes fondamentaux de la théorie des fonctions', in Molk (ed.), *Encyclopédie des sciences mathématiques pures et appliquées*. Paris: Gauthiers-Villars, ii(1). 1–112.

MOORE, G. H. (1982), *Zermelo's Axiom of Choice: Its Origins, Development and Influence*. Berlin: Springer.

MULLIGAN, K. (1985), ' "Wie die Sachen sich zueinander verhalten" Inside and Outside the *Tractatus*', in B. F. McGuinness and A. G. Gargani (eds.), *Wittgenstein and Contemporary Philosophy*. Pisa: ETS, 145–74.

—— SIMONS, P., and SMITH, B. (1985), 'Truth-Makers', *Philosophy and Phenomenological Research* 44: 287–321.

MYHILL, J. (1974), 'The Undefinability of the Set of Natural Numbers in the Ramified *Principia*', in G. Nakhnikian (ed.), *Bertrand Russell's Philosophy*. London: Duckworth, 19–27.

NEDO, M., and RANCHETTI, M. (1983), *Ludwig Wittgenstein: Sein Leben in Bilder und Texten*. Frankfurt: Suhrkamp.

NELSON, E. (1986), *Predicative Arithmetic*. Princeton, NJ: Princeton University Press.

—— (1994), 'Taking Formalism Seriously', in D. Prawitz, B. Skyrms, and D. Westerståhl (eds.), *Logic, Methodology and Philosophy of Science ix*. Amsterdam: Elsevier, 571–7.

NOË, A. (1994), 'Wittgenstein, Phenomenology and What it Makes Sense to Say', *Philosophy and Phenomenological Research* 54: 1–42.

PARIKH, R. (1971), 'Existence and Feasibility in Arithmetic', *Journal of Symbolic Logic* 36: 494–508.

PARSONS, C. (1983), 'The Impredicativity of Induction', in L. S. Cauman et al. (eds.), *How Many Questions? Essays in Honor of Sidney Morgenbesser*. Indianapolis: Hackett, 132–53.

PEARS, D. (1987), *The False Prison*, i. Oxford: Oxford University Press.

PEIRCE, C. S. (1933), 'On the Algebra of Logic', in C. Hartshorne and P. Weiss (eds.), *Collected Papers of Charles Sanders Peirce*, iii: *Exact Logic*. Cambridge, Mass.: Harvard University Press, 210–49.

PITCHER, G. (1964), *The Philosophy of Wittgenstein*. Englewood Cliffs, NJ: Prentice-Hall.

POINCARÉ, H. (1906), 'Les Mathématiques et la logique', in *Revue de métaphysique et de morale* 14: 294–317.

—— (1952a), *Science and Hypothesis*. New York: Dover.

—— (1952b), *Science and Method*. New York: Dover.

—— (1963), *Mathematics and Science: Last Essays*. New York: Dover.

PRINGSHEIM, A. (1899), 'Grundlagen der allgemeinen Funktionenlehre', in H. Burkhardt and W. F. Meyer (eds.), *Encyklopädie der mathematischen Wissenschaften mit Einschluss ihrer Anwendungen*. Leipzig: Teubner, IIa1. 1–41.

PUTNAM, H. (1981), *Reason, Truth and History*. Cambridge: Cambridge University Press.

QUINE, W. V. (1962), *Methods of Logic*, 2nd edn. London: Routledge & Kegan Paul.

RAMSEY, F. P. (1923), 'Critical Notice: *Tractatus Logico-Philosophicus* by Ludwig Wittgenstein', *Mind* 32: 465–78.

—— (1978), *Foundations*, ed. D. H. Mellor. London: Routledge & Kegan Paul.

—— (1990), *Notes on Philosophy, Probability and Mathematics*, ed. M.-C. Galavotti. Naples: Bibliopolis.

—— (1991), *On Truth*, ed. N. Rescher and U. Majer. Dordrecht: Kluwer.

RHEES, R. (ed.) (1984), *Recollections of Wittgenstein*. Oxford: Oxford University Press.

RICHARD, J. (1967), 'The Principles of Mathematics and the Problem of Sets', in van Heijenoort (1967: 142–4).

RIEMANN, B. (1953), *Gesammelte mathematische Werke und wissenschaftlicher Nachlass*, ed. H. Weber. New York: Dover.

ROBINSON, R. M. (1947), 'Primitive Recursive Functions', *Bulletin of the American Mathematical Society* 53: 925–42.

RUSSELL, B. (1906), 'On Some Difficulties in the Theory of Transfinite Numbers and Order Type', *Proceedings of the London Mathematical Society*, 2nd ser., 4: 29–53.

—— (1910), 'La théorie des types logiques', *Revue de métaphysique et de morale* 18: 263–301.

—— (1919), *Introduction to Mathematical Philosophy*. London: Allen & Unwin.

—— (1931), 'Critical Notice on *The Foundations of Mathematics and Other Logical Essays*. By Frank Plumpton Ramsey', *Mind* 40: 476–82.

—— (1936), 'The Limits of Empiricism', *Aristotelian Society Proceedings*, n.s., 36: 131–50.

—— (1968), *The Autobiography of Bertrand Russell: 1914–1944*. London: Allen & Unwin.

—— (1984), *Theory of Knowledge*. London: Allen & Unwin.

—— (1985), *The Philosophy of Logical Atomism*, ed. D. Pears. La Salle, Ill.: Open Court.

—— (1986), *Mysticism and Logic*. London: Allen & Unwin.

—— (1989), *The Analysis of Mind*. London: Routledge.

—— (1993), *Our Knowledge of the External World*. London: Routledge.

Ryle, G. (1990), ' "If", "So", and "Because" ', in *Collected Essays*, ii. Bristol: Thoemmes, 234–49.

Sahlin, N.-E. (1990), *The Philosophy of F. P. Ramsey*. Cambridge: Cambridge University Press.

—— (1995), 'On the Philosophical Relations between Ramsey and Wittgenstein', in Hintikka and Puhl (1995: 150–63).

Scott, D. (1970), 'Outline of a Mathematical Theory of Computation', Technical Monographs PRG-2, Programming Research Group, 8–11 Keble Road, Oxford.

Scott, D. S., and Tarski, A. (1958), 'The Sentential Calculus with Infinitely Long Expressions', *Colloquium Mathematicae* 6: 165–70.

Shanker, S. G (ed.), 1986, *Ludwig Wittgenstein: Critical Assessments*, iii. London: Croom Helm.

—— (1987), *Wittgenstein and the Turning Point in the Philosophy of Mathematics*. London: Croom Helm.

—— (1987), 'Wittgenstein's Remarks on the Significance of Gödel's Theorem', in Shanker (ed.), *Gödel's Theorem in Focus*. London: Croom Helm, 155–256.

—— (1993), 'Wittgenstein versus Russell on the Analysis of Mind', in A. Irvine and G. A. Wedeking (eds.), *Russell and Analytic Philosophy*. Toronto: University of Toronto Press, 210–42.

Sheffer, H. M. (1913), 'A Set of Five Independent Postulates for Boolean Algebras, with Applications to Logical Constants', *Transactions of the American Mathematical Society* 14: 481–8.

Sinaceur, H. (1991), *Corps et modèle: Essai sur l'histoire de l'algèbre réelle*. Paris: Vrin.

Skolem, T. (1967a), 'Some Remarks on Axiomatized Set Theory', in van Heijenoort (1967: 291–301).

—— (1967b), 'The Foundations of Elementary Arithmetic Established by Means of the Recursive Mode of Thought, Without Use of Apparent Variables Ranging over Infinite Domains', in van Heijenoort (1967 303–33).

SMORYNSKI, C. (1988), 'Hilbert's Programme', Dept. of Philosophy, University of Utrecht.

SOAMES, S. (1983), 'Generality, Truth-Functions, and Expressive Capacity in the *Tractatus*', *Philosophical Review* 92: 573–89.

SPIEGELBERG, H. (1968), 'The Puzzle of Wittgenstein's *Phänomenologie* (1929–?)', *American Philosophical Quarterly* 5: 244–56.

STATMAN, R. (1974), *The Structural Complexity of Proofs*. Dissertation, Stanford University.

STEINER, M. (1975), *Mathematical Knowledge*. Ithaca, NY: Cornell University Press.

STEINITZ, E. (1930), *Algebraische Theorie der Körper*, ed. R. Baer and H. Hasse. Berlin: Springer, repr. New York: Chelsea, 1950.

STERN, D. (1991), 'The "Middle Wittgenstein": From Logical Atomism to Practical Holism', *Synthese* 87: 203–26.

—— (1994), *Wittgenstein on Mind and Language*. Oxford: Oxford University Press.

SUNDHOLM, G. (1992), 'The General Form of the Operation in Wittgenstein's *Tractatus*', *Grazer Philosophische Studien* 42: 57–76.

—— (1994), 'Existence, Proof and Truth-Making: A Perspective on the Intuitionistic Conception of Truth', *Topoi* 13: 117–26.

TAIT, W. W. (1968), 'Constructive Reasoning', in B. van Rootselaar and J. F. Stall (eds.), *Logic, Methodology and Philosophy of Science*, ii. Amsterdam: North-Holland, 185–99.

—— (1981), 'Finitism', *Journal of Philosophy* 78: 524–46.

TARSKI, A. (1951), *A Decision Method for Elementary Algebra and Geometry*. Berkeley, Calif.: University of California Press.

—— (1965), *Introduction to Logic and the Methodology of Deductive Sciences*. Oxford: Oxford University Press.

—— (1983), *Logic, Semantics, Metamathematics*, 2nd edn. Indianapolis: Hackett.

TROELSTRA, A. N. (1982), 'On the Origin and Development of Brouwer's Concept of Choice Sequence', in Troelstra and van Dalen (1982: 465–86).

—— and VAN DALEN, D. (eds.) (1982), *The L. E. J. Brouwer Centenary Symposium*. Amsterdam: North-Holland.

—— —— (1988), *Constructivism in Mathematics: An Introduction*, i. Amsterdam: North-Holland.

TROELSTRA, A. S. (1977), *Choice Sequences: A Chapter in Intuitionistic Mathematics*. Oxford: Oxford University Press.

VAN BENDEGEM, J. P. (1987), *Finite, Empirical Mathematics: Outline of a Model*. Ghent: Rijksuniversiteit.

VAN DANTZIG, D. (1955), 'Is $10^{10^{10}}$ a Finite Number?' *Dialectica* 9: 273–7.

VAN DEN DRIES, L. (1988), 'Alfred Tarski's Elimination Theory for Real Closed Fields', *Journal of Symbolic Logic* 53: 7–19.

VAN DER WAERDEN, B. L. (1931), *Moderne Algebra*. Berlin: Springer. English translation, *Modern Algebra*. New York: Frederick Ungar, 1949.

VAN HEIJENOORT, J. (ed.) (1967), *From Frege to Gödel: A. Sourcebook in Mathematical Logic, 1879–1931*. Cambridge, Mass.: Harvard University Press.

VAN PEURSEN, C. A. (1970), *Ludwig Wittgenstein. An Introduction to His Philosophy*. New York; Dutton.

VAN STIGT, W. P. (1990), *Brouwer's Intuitionism*. Amsterdam: North-Holland.

VLADUT, S. G. (1991), *Kronecker's Jugendtraum and Modular Functions*. New York: Gordon & Breach.

VON PLATO, J. (1993), *Creating Modern Probability*. Cambridge: Cambridge University Press.

VON WRIGHT, G. H. (1971), 'Wittgenstein's Views on Probability', in *Wittgenstein et le problème d'une philosophie de la science*. Paris: CNRS, 113–33.

—— (1984), *Truth, Knowledge, and Modality*. Oxford: Blackwell.

WAISMANN, F. (1951), *Introduction to Mathematical Thinking*. London: Hafner.

—— (1968), *How I See Philosophy*. London: Macmillan.

—— (1982), *Lectures on the Philosophy of Mathematics*, ed. W. Grassl. Amsterdam: Rodopi.

—— (1986), 'The Nature of Mathematics: Wittgenstein's Standpoint', in Shanker (1986: 60–7).

WANG, H. (1958), 'Eighty Years of Foundational Studies', *Dialectica* 12: 466–97.

—— (1961), 'Process and Existence in Mathematics', in Y. Bar Hillel et al. (eds.), *Essays on the Foundations of Mathematics dedicated to A. A. Fraenkel on His Seventieth Anniversary*. Jerusalem: Magnes, 328–51.

WATSON, A. G. D. (1938), 'Mathematics and Its Foundations', *Mind* 47: 440–51.

WEBER, H. (1893), 'Leopold Kronecker', *Jahresbericht der deutschen Mathematiker-Vereinigung* 2: 5–31.

WEIL, A. (1976), *Elliptic Functions according to Eisenstein and Kronecker*. Berlin: Springer.

WEYL, H. (1940), *The Algebraic Theory of Numbers*. Princeton, NJ: Princeton University Press.

—— (1959), *Philosophy of Mathematics and Natural Sciences*. Princeton, NJ: Princeton University Press.

—— (1967), 'Comments on Hilbert's Second Lecture on the Foundations of Mathematics', in van Heijenoort (1967: 482–4).

—— (1968), *Gesammelte Abhandlungen*, ed. K. Chandrasekharan. Berlin: Springer.

—— (1987), *The Continuum: A Critical Examination of the Foundations of Analysis*. Kirksville, Mo.: Thomas Jefferson University Press.

WHITE, R. (1979), 'Wittgenstein on Identity', *Aristotelian Society Proceedings*, n.s., 78: 157–74.

WRIGLEY, M. (1986), 'Wittgenstein on Inconsistency', in Shanker (1986: 347–59).

—— (1987), 'Wittgenstein's Early Philosophy of Mathematics'. Dissertation, University of California at Berkeley.

—— (1991), 'Infinity in the *Tractatus*', MS.

—— (1993), 'The Continuity of Wittgenstein's Philosophy of Mathematics', in K. Puhl (ed.), *Wittgenstein's Philosophy of Mathematics: Proceedings of the 15th International Wittgenstein-Symposium*, ii. Vienna: Hölder-Pichler-Tempsky, 73–83.

—— (1995), 'Wittgenstein, Ramsey and the Infinite', in Hintikka and Puhl (1995: 164–70).

WRIGHT, C. J. G. (1980), *Wittgenstein on the Foundations of Mathematics*. London: Duckworth.

—— (1983), *Frege's Conception of Numbers as Objects*. Aberdeen: Aberdeen University Press.

—— (1985), 'Skolem and the Skeptic', *Aristotelian Society Proceedings*, supplementary vol. 59: 117–37.

—— (1993), 'Strict Finitism', in *Realism, Meaning and Truth*. Oxford: Blackwell, 107–75.

ZERMELO, E. (1904), 'Proof That Every Set Can Be Well-Ordered', in van Heijenoort (1967: 139–41).

INDEX